Hypersonic Flows for Reentry Problems

Volume I: *Survey Lectures and Test Cases for Analysis*

J.-A. Désidéri R. Glowinski J. Périaux (Eds.)

Hypersonic Flows for Reentry Problems

Volume I:
Survey Lectures and Test Cases for Analysis

Proceedings of a Workshop
Held in Antibes, France, 22–25 January 1990

With 114 Figures

Springer-Verlag
Berlin Heidelberg New York London Paris
Tokyo Hong Kong Barcelona Budapest

Dr. Jean-Antoine Désidéri
INRIA, Centre de Sophia Antipolis,
2004 Route des Lucioles, F-06561 Valbonne Cedex, France

Professor Dr. Roland Glowinski
University of Houston, Texas, Department of Mathematics,
4800 Calhoun Road, University Park, Houston, TX 77004, USA

Dr. Jacques Périaux
Aerodynamics Department, Dassault Aviation,
78 Quai Marcel Dassault, F-92214 Saint Cloud, France

ISBN-13:978-3-642-84582-6 e-ISBN-13:978-3-642-84580-2
DOI: 10.1007/978-3-642-84580-2

Library of Congress Cataloging-in-Publication Data Hypersonic flows for reentry problems: proceedings of a workshop, held in Antibes, France, 22–25 January 1990/J.-A. Désidéri, J. Périaux, R. Glowinski, editors. p. cm. Includes bibliographical references and indexes. Contents: v. 1. Survey lectures and test cases for analysis – v.
ISBN-13:978-3-642-84582-6 1. Space vehicles-
Atmospheric entry-Congresses. 2. Aerodynamics, Hypersonic-Congresses. 3. Space vehicles-Hypersonic-Problems, exercises, etc. 4. Aerodynamics, Hypersonic-Problems, exercises, etc. I. Désidéri, J.-A. (Jean-Antoine), 1951- II. Périaux, Jacques. III. Glowinski, R. TL 1060.H96 1991 629.4' 15-dc20 91-32792

Softcover reprint of the hardcover 1st edition 1991

Typesetting: Camera ready by author

55/3140-543210 – Printed on acid-free paper

Preface

One of the most challenging problems of modern engineering is undoubtedly the prediction of hypersonic flows around space vehicles in reentry conditions. Indeed, the difficulties are numerous: first of all, these flows are very difficult to model, since very complex physical and chemical phenomena take place during the reentry phase; secondly, temperature, velocity and enthalpy are very high and densities are very low, making the reentry process very difficult to reproduce in ground-based experiments.

The past three decades have seen important efforts in computational fluid dynamics relying on the use of supercomputers to simulate these very complicated flows. The numerical simulation based on imperfect models and methods which were essentially designed for transonic and supersonic flows has still a long way to go in order to be able to predict these hypersonic reentry flows very accurately.

This situation has motivated very strong international cooperative efforts with, as the most visible consequences, the Europe/United States Short Courses on Hypersonics, which were held in Paris, in 1987 [1,2], Colorado Springs in 1989 [3], and Aachen in 1990 [3].

The workshop on Hypersonics whose results are presented and analysed in these volumes is also a direct consequence of this international cooperation. This scientific event was an initiative of P. Perrier, Head of the Theoretical Aerodynamics Department of DASSAULT AVIATION, who played a key role in the identification of the critical problems and the realisation of experiments, within the Hermès R&D program framework.

During this workshop, which took place in Antibes, France, in January 1990, about 200 engineers and scientists from Europe, the United States, Japan and Australia presented results associated with the solution of 8 test problems. These 8 problems were carefully selected by an International Committee as representative of real life difficulties existing during the reentry of space vehicles, such as the European Hermès and the US Orbiter.

Actually, during the workshop, the technical presentations of the test problem results were completed by 10 invited lectures given by leading scientists on topics related to modeling, experimental aspects and numerical simulations of hypersonic flows.

Volume 1 contains the invited lectures and descriptions of the test problems. Volume 2 is in a sense more technology oriented. For each problem a synthesis of the results obtained by the various workshop participants is followed by the presentation of the experimental and computational contributions.

The preparation of this workshop and of the proceedings has been quite a formidable task.

We would like to express our particular thanks to R. Abgrall, A. Dervieux, M. Mallet and B. Stoufflet for their continuous and friendly scientific assistance in the preparation and the development of the workshop and the realisation of these volumes.

The development of the workshop was a success due to the very close cooperation of the following people whose talents cover very large and impressive spectra in science and engineering, organization and management, and last but not least, fund raising abilities among others:

Steering Committee

S. Bogdonoff (USA); R. Brun (France); C. Cercignani (Italy); A. Dervieux (France); C. Dujarric (ESA); R. Glowinski (USA/France); E. Krause (Germany); B. Mercier (France); K. Morgan (UK); P. Morice (France); H. Neunzert (Germany); R. Pellat (France); J. Périaux (France); P. Perrier (France); O. Pironneau (France); A. Rizzi (Sweden).

Scientific Committee

R. Abgrall (INRIA Sophia Antipolis, France); M.O. Bristeau (INRIA Rocquencourt, France); A. Dervieux (INRIA Sophia Antipolis, France); J.-A. Désidéri (INRIA Sophia Antipolis and GAMNI/SMAI, France); G. Durand (CNES Evry, France); L. Fezoui (INRIA Sophia Antipolis, France); R. Glowinski (Univ. of Houston, USA and INRIA, France); F. Gropengiesser (Univ. of Kaiserslautern, Germany); M. Imbert (CNRS Marseille, France); P. Le Tallec (Univ. Paris Dauphine and GAMNI/SMAI, France); M. Mallet (Dassault Aviation and GAMNI/SMAI, France); L. Marraffa (ONERA, France); M. Pandolfi (Politecnico di Torino, Italy); J. Périaux (Dassault Aviation and GAMNI/SMAI, France); Ph. Rostand (INRIA Rocquencourt, France); B. Stoufflet (Dassault Aviation, France); Ph. Vancamberg (Dassault Aviation, France); D. Vandromme (Univ. of Rouen, France); S.A. Waiter (Dassault Aviation, France); D. Zeitoun (CNRS Marseille, France).

Local Organizing Committee

Th. Bricheteau (INRIA, France); S. Gosset (INRIA, France); C. Juncker (INRIA Sophia Antipolis, France); P. Rousseaux (Dassault Aviation, France); G. Thaler (SMAI, France); N. Torossian (Dassault Aviation, France)

Evaluation Committee

Th. Alziary de Rocquefort (Univ. of Poitiers, France); S. Bogdonoff (Princeton, USA); M. Borsi (Aeritalia, Italy); G. Durand (CNES, France); A. Dervieux (INRIA, France); J.-A. Désidéri (INRIA and GAMNI/SMAI, France); P. Fauchais (Univ. of Limoges, France); W. Goodrich (AGARD, USA); M. Green (NASA-Ames, USA); A. Kumar (NASA-Langley, USA); E. Krause (RWTH Aachen, Germany); F. Gropengiesser (Univ. of Kaiserslautern, Germany); J.K. Harvey (Imperial College, UK); W. Kordulla (DLR, Germany); M. Linde (FFA, Sweden); P. Morice (ONERA, France); J. Muylaert (ESTEC, The Netherlands); M. Pandolfi (Politecnico di Torino, Italy); J. Périaux (Dassault Aviation and GAMNI/SMAI, France), P. Perrier (Dassault Aviation, France); O. Pironneau (Univ. of Paris VI and INRIA, France); A. Rizzi (KTH, Sweden); B. Stoufflet (Dassault Aviation, France); H. Viviand (Principia, France); S.A. Waiter (Dassault Aviation, France).

The organizers would like to acknowledge the quality of the invited and contributed presentations and also the Chairmen of the various sessions and of the Evaluation Committee since they did an outstanding job under considerable pressure to keep the workshop on schedule.

The organizers would like to address their warmest thanks to those institutions and companies that have supported the workshop, namely AGARD, CIRA, CNES, CRAY, DASSAULT AVIATION, ESA.

Special thanks are due to the Service des Relations Extérieures of INRIA, which masterminded a more than pleasant week on the French Riviera that most participants enjoyed tremendously, for its help and assistance in the preparation of the workshop.

The editors would like to thank Springer-Verlag and more particularly Professor W. Beiglböck for kindly agreeing to publish these highly specialized proceedings.

Last, but not least, the editors would like to acknowledge Sabine Landgraf and Sylviane Gosset for their help and patience with us during the processing of the full manuscript.

We hope that you, the reader, will find these volumes a classic reference that will be used frequently in the years to come.

December 1990

Jean-Antoine Désidéri, INRIA Sophia Antipolis and GAMNI/SMAI
Roland Glowinski, University of Houston and INRIA
Jacques Périaux, Dassault Aviation and GAMNI/SMAI.

References

[1] J.J. Bertin, R. Glowinski, J. Périaux, Editors, Proceedings of the First Joint Europe/US Short Course in Hypersonics, Paris, France, Dec. 7–11, 1987, Defining the Hypersonic Environment, Hypersonics, Volume I, Progress in Scientific Computing, Birkhäuser Boston, 1989.

[2] J.J. Bertin, R. Glowinski, J. Périaux, Editors, Proceedings of the First Joint Europe/US Short Course in Hypersonics, Paris, France, Dec. 7–11, 1987, Computation and Measurement of Hypersonic Flows, Hypersonics, Volume II, Progress in Scientific Computing, Birkhäuser Boston, 1989.

[3] J. Ballmann, J.J. Bertin, J. Périaux, Editors, Proceedings of the Second and Third Joint Europe/US Short Courses in Hypersonics, Colorado Springs, USA, January 16–20, 1989, and Aachen, Germany, October 1–5, 1990, Progress in Scientific Computing, Birkhäuser Boston, to appear.

Contents

Problems for Analysis

INTRODUCTION

When advanced hypersonic projects came to existence in Europe, mainly with the HERMES proposal, it was obvious that technological and scientific levels had to be enhanced. The availability of sufficient knowledge in hypersonic is mandatory before any realistic time schedule for development can be ascertained. Low cost in development relies now on a numerical simulation - experimental test and flight test progressive methodology : use of a mix of the best of each method as soon as possible is a key to success of advanced aerospace programs.

The past efforts made in the sixties in Europe and U.S. and more recent work on U.S. side, for Shuttle Orbiter design, do not seem sufficient for giving confidence in design of advanced reentry vehicles. Significant improvements in experiments in physical and chemical sciences have been accomplished recently ; but moreover the general methodology of design has to take into account the major role played now by CFD in new project design ; quality of CFD evidently relies on correctness in physical and chemical modelling as well as on the numerical modelling of the PDE problem to be solved. The rational approach of modelling with computer codes embodies capabilities of rebuilding complex flows impossible to be rebuilt in ground test facilities ; but it needs to be comforted by an elementary experimental validation on a limited number of basic models. It is clear that such effort has not been pushed in the past ; validation is now recognized as a necessary step in the evaluation of level of technology, therefore it is a part of any feasibility study of new hypersonic projects.

Thanks to CNES and ESA some support to the necessary effort in experimental and theoretical work has been brought in Europe as part of the HERMES project preliminary studies; it has been very useful for all the research or industrial community working on hypersonic. The success of HERMES Research for Development (RfD) program, covering large number of scientific organizations in Europe encouraged after 3 years an attempt to a confrontation of the computation and experiments on carefully selected problems.

Opening such a confrontation to all world scientists would allow a more general evaluation of the uncertainties in CFD rebuilding of the physics and help to define guidelines for future work.

A Scientific Committee has selected a few number of problems of gradual difficulties. Schematic problems were retained ; they are basically helpful to check quality of flow rebuilding from scientific and industrial points of view. Moreover selection of basic problems helps to better evaluate it in an open international context.

In January 1990, the first international workshop on hypersonic flows for reentry vehicles was a successful event by the number of participants and many encouraging results ; unequal difficulties were encountered by participants in the tests cases. It proves the achievements done in CFD as well as the large improvement remaining to be done. Reference tests cases have proven their pertinence by stressing a lot of unsolved or poorly solved problems with existing computers ; some evidence of the origin of the problems encountered are either in numerical scheme, or on discretization and also on physical relevant modelling.

The material collected in this book takes advantage of revision of many contributions after the workshop, first theory vs experiment confrontation, has taken place ; some additional contributions appeared after this event but were retained due to their added value. However it appears that better results could be obtained with more regular or on more refined meshes on the selected cases. Consequently a Part II Workshop will take place in 1991 and will complement the work presented here.

The workshop was made possible by the active cooperation of CNES-ESA through DASSAULT AVIATION, INRIA and GAMNI/SMAI. It would have been difficult to exploit all the material selected before the event presented in sessions without the help of all the Members of Scientific Reviewing Committee and and the chairmen of the sessions . Preliminary synthesis done for the final session of the workshop have been carefully revisited with last materials and are presented before the experimental-numerical comparisons. Some gratitude has to be expressed to the authors of the synthesis as well as to the authors of contributions.

Particular thanks are due to Doctor J.PERIAUX of DASSAULT AVIATION and Doctor J.A. DESIDERI, INRIA who help to materialize the target of the workshop and of these proceedings in cooperation with the authors , session chairmen and participants : all contributed to the success of the Workshop.

INVITED LECTURES

Workshop on Hypersonic flows for reentry problems
January 22 - 25th 1990 (Antibes)

Inaugural Address

J. CARPENTIER

Before introducing the subject of this workshop, I would like to congratulate INRIA and GAMNI-SMAI for their joint organization of this event. The purpose of this workshop is to bring together experts actively working in Hypersonics and to ask them to discuss about the knowledge of flows at high Mach numbers and large angles of attack.

This initiative is particularly appropriate in the perspective of new European and American space projects.

Before discussing the specific field of Hypersonics, it is worth considering the ways of progress concerning aerodynamics, in general.

The exploration of the high speed flights began with a joint approach between wind tunnel tests and experimental vehicles (viewgraph 1). This joint approach has been very sccessful. Among the many examples of these success, we can quote, in the States, the supersonic flights of the Bell X1 in 1947 and the hypersonic flights of the North American X15 which attained M6 in 1961. In Europe also, each aeronautical nation led and experimental investigation, more or less tightened to the realization of prototypes. Among the forerunners, we could mention the Dassault OURAGAN and the Nord-Aviation Griffon. This

Introduction to the workshop

J. CARPENTIER

* *Wind Tunnel Limitations*
* *Experimental Vehicles*
* *Computational Fluid Dynamics*
* *The necessity of a joint approach :*
  - *CFD*
  - *Ground experimentation*
  - *Flight experimentation*
* *Conclusions.*

LIMITE DES POSSIBILITES DE TRAITEMENT

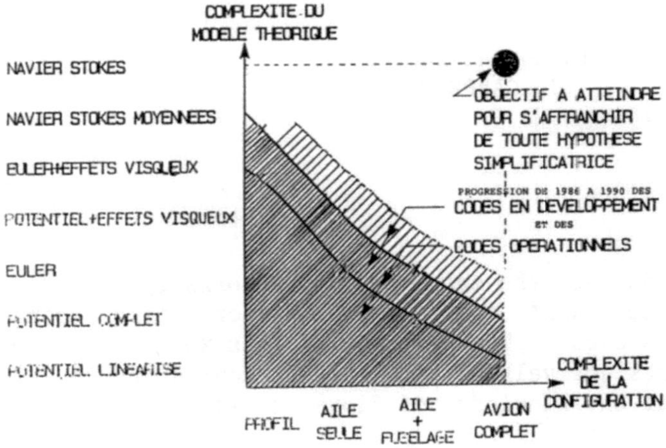

one was a turboramjet aircraft, and with this new type of engine, the Griffon attained M2,2 in 1957 (climbing). An ONERA experimental missile, STATALTEX powered by ramjet flew at M5 (38.000 m) in 1962.

But the apparition of numerical computers has completely renewed the possibility of computation.

For instance, this chart (viewgraph 2) shows the progress of Computational Fluid Dynamics for the determination of the flow in the subsonic field on the X axis is presented the configuration complexity. The simplest is an airfoil which needs only 2D computations. The most complex is a complete aircraft.

On the Y axis, you can see different computation methods from the simplest potential methods to the complete Navier-Stokes method.

It is possible to calculate the flow around an aircraft with the simplest method. It is also possible to calculate the flow around an airfoil with a complex method. But this is not the case for a complete aircraft.

The limit of the domain covered by CFD for subsonic flow is moving from year to year. You can see here the progress from 1986 till now.

But the ultimate target is not to-morrow.

And it's the same if not worst for higher speeds and especially for hypersonics where we have to take account of reacting flows.

Wind Tunnel limitations

	Mach	I	W.T. limitations
Transport aircraft – subsonic – SST	 0.9 2–3	 * **	 M ~ 1 Re Turbulence and noise
Combat aircraft	2–3	**	High α I
Re–entry vehicles	25		Enthalpy and real gas simulation I and measurement I
Hypersonic airplanes	5–25	***	
Horizontal launchers	25	***	

I = Propulsion Integration

It is the main purpose of this workshop to precise the knowledge of the CFD possibilities for different configurations of hypersonic bodies from the simplest one : cone, 2D ramp, to more representative bodies such as generic shapes of aerospace vehicles. These different configurations could be positioned along the X axis.

Along the Y axis would be placed different computation methods from the simplest to the more complexe one.

So, the way to the ultimate target which could be the use of CFD alone is still a long way. The "numerical wind-tunnel" is not for to-morrow. Physical wind-tunnels have still several decades -at least- for their future. But they have to be designed or to be modified to the new approach.

At first, we can see some limitations of existing wind-tunnels (viewgraph 3) :
- for transport aircrafts,
- for combat aircrafts,
- for reentry vehicles...

So it's necessary to extend the possibilities of wind-tunnels. But the main effort is to be focused on the validation of numerical codes in wind-tunnels.

I give you some examples of what we are doing at ONERA conventional wind-tunnels S2Ch ... S5Ch and for blowdown WT R2Ch - R3Ch which are now equipped with precise instrumentation devices (viewgraphs 4 and 5). The code validation is also a reason why we develop our T2 CERT which will be very useful for laminarity research - and our F2 CFM which has been specifically designed for precise optical measurements of 3D flows.

Validation in ONERA Wind-Tunnels

Wind-Tunnel	Characteristics	Validations	Measurements
S2 Ch Chalais-Meudon	Ø 3m 120m/s	Helicopter rotors Ellipsoids	Forces-Moments Pressures Laser Velocimetry
S3 Ch Chalais-Meudon	0.8mx0.8m 0.3<M<1.2	Afterbodies Wing-nacelle integration	Forces-Moments Pressures 3D Laser Velocim
S5 Ch Chalais-Meudon	0.3mx0.3m 0.1<M<2 1.4<M<3.15	Blade cascades Air intakes	Pressures 3D Velocim. prob Laser Velocimetry

Validation in ONERA Wind-Tunnels

Wind-Tunnel	Characteristics	Validations	Measurements
R2 Ch - R3 Ch Chalais-Meudon	Ø 0.35m T_i =1100K M3/4/5/6/7/10 10 à 30s	Shock-boundary layer Launcher afterbodies Re-entry kinetic heating	Forces-Moments Pressures Flux Laser Velocimetry
T2 CERT Toulouse	0.4mx0.4m 90<T_i<300K 3 bar 0.6<M<0.9	Airfoils Laminarity-Transition Buffeting	Pressures Wakes Laser Velocimetry
F2 CFM Fauga-Mauzac	1.4mx1.8m 100 m/s	Ellipsoïds 3D generic shapes	Pressures Wakes 3D Laser Velocim.

And going back to the field of hypersonics, we have added to our existing facilities such as R2Ch - R3Ch and S4 Modane, an entirely new one F4.

This hot shot arc-heated wind-tunnel will be an excellent tool to validate codes for the european Hermes space plane.

So the race between CFD and wind-tunnels will lead to new types of experimental facilities. They will be designed to afford :
- very well defined flow conditions,
- very precise, fast, and non intrusive measurements.

They will be easy to operate and to maintain, their main objectives will be to realize tests on generic shapes easy to calculate, in order to validate computation codes, and this, within very precise physical conditions.

Thus, there will not be a real challenge between the CFD or numerical wind-tunnels on one hand, and physical wind-tunnels on the other hand. These two types of approaches are the two parts of a joint approach.

This workshop, with the very high level of the participants all engaged in an active work in Hypersonics, will be a milestone in this powerful approach.

The results will be :
- a better knowledge of hypersonic flows,
- a whole set of precise tools giving access to the design of very efficient hypersonic planes.

COMMENTS ON EXPERIMENTS FOR COMPUTATIONAL VALIDATION FOR FLUID DYNAMIC PREDICTIONS

By

Seymour M. Bogdonoff
Department of Mechanical and Aerospace Engineering
Gas Dynamics Laboratory
Princeton University
Princeton, NJ 08544 USA

1. Introduction

Over the past two decades, there have been extraordinary advances in the techniques and technology of computation and a many-fold increase in the capacity and speed of computers. For problems of fluid mechanics, if the equations which fully describe the flow are known, the problems of computation are primarily the discretization processes. The problems are characterized by questions of grid configuration and density, and computational errors (rounding and averaging). If exact solutions of the equations are known, then the computational capability and accuracy can be evaluated by direct comparison. There are, however, very few such situations in the field of the aerodynamic design of vehicles.

For many problems of fluid mechanics, the direct evaluation of the computation requires some check against experiments. Although this has been done, in the early stages of computation, in a rather cursory manner, recent more direct dependence on computation to provide "design" information requires a much more thorough comparison of experiments and computation to evaluate the capabilities and accuracy of computations used for design.

Although this process has been going on for some time, it has only recently come into major focus. A series of papers have been written to provide guidance for such "computational validation" processes, for example, Refs. 1-3. The present paper focuses on the author's experience with such processes over the past two decades, where detailed experiments and computations, <u>in combination</u>, have provided the basis for new understanding of complex flows. Experiments performed for the usual research, development, and engineering investigations are seldom, if ever, totally adequate for computational validation. It is the special requirements for the experiments, and the interactions with computations, which are the thrust of the subject paper.

2. *Framework for Experiments for Computational Validation*

a) *Initial phase*

Most of the computational validation carried out in the past decade has consisted in the direct comparison of several computational methods and refinements to some specific experiments. The experiments were, in most cases, not specifically designed to test the computation but rather to explore a particular physical phenomena. The experiments were carried out under whatever test conditions seemed to be appropriate, and with the instrumentation available. Although valuable information can be obtained from such comparisons, the results are, at best, qualitative and provide a early test of computation, grid refinement, and other techniques. The results, however, do not give a "validation" check adequate for the design of arbitrary vehicles under all flight conditions.

b) *Interactive phase*

The studies of Section a) were usually carried out by an experimentalist, with some discussion with CFD staff. However, the author believes that future studies must be designed specifically to test computations in ways which have not generally been applied in Section a). There are, for example, few studies which have determined what parameters are sensitive in any physical flow, and what specific parameters are required, with what accuracy, to reasonably describe the flow. CFD has the capability to test sensitivity, and should be used to determine what parameters are sensitive to assumptions included in CFD and in the experiments. Before experiments are made, CFD should determine the grid requirements for the computation which give, as best they know, adequate results, and the sensitivity of the phenomena to initial conditions or perturbations of the initial conditions should be specified from extensive computation.

c) The design of experiments for CFD validation

The experiments should be designed to stress the computation, with parameters that are needed by the designer. The initial conditions have to be known in great detail, and the boundary conditions have to be fully described. The CFD activities of Section b) should specify to the experimentalist the requirements for the test. General observations such as "the flow is two-dimensional" is not adequate. Although the geometry may be two-dimensional, the flowfield, in almost all cases, has some three-dimensional structure. In supersonic flow, the inclusion of small subsonic regions will almost always give considerable experimental and computational difficulty. The experiments should provide more than one specific set of measurements. Flowfield details, as best can be measured, are required to verify computed flowfield structures, and surface conditions require multiple measurements. Surface visualization, which has been used extensively, can be misleading without flowfield information and gives no indication of flowfield unsteadiness. Surface pressure distributions are determined, in large part, by the flowfield, while skin friction or heat transfer is primarily determined by the local conditions close to the wall, areas which are not, in general, fully probed in experiments nor totally modeled in computation. The question of flowfield unsteadiness is, in most cases, an assumption which has not been borne out in many detailed measurements. Finally, there is the question of whether the wind tunnel flow is, in any way, a true simulation of the conditions which would be in flight. At the least, the full wind tunnel characteristics must be included for specific inclusion in the computations, while the question of whether the wind tunnel duplicates flight is another question which must be answered for the designer of vehicles.

d) Range

No single point experiment, that is, one geometry at one test condition, is really adequate, even though it may be carried out in great detail. Such an experiment is very good to examine the computational capability for details, but the variation of parameters is a much more critical test of computation. Variation of initial conditions, for example, boundary layer, Mach number, wall temperature, provide real tests of the ability of computation to provide information required by the designer. Variable geometry, for example, fin angle or sweep angle at variable Mach numbers, provides a critical test of computations. Computaton may appear to give good results under one condition, but may not be able to be used generally over a wide range.

e) Instrumentation

Seldom, if ever, can the experiments measure all of the parameters which are needed to fully validate computations. Whatever can be measured, should be measured, with redundancy as a real goal. The continuing development of new instrumentation, which will increase accuracy, resolution, and frequency response, will provide, in time, new information which must be continuously evaluated in computational validation. The most critical test, in the author's experience, would be the experimental determination of specific streamlines in complex three-dimensional flows, and a comparison of these streamline paths to computation. This is, at the moment, not a capability which has been demonstrated, but appears to be possible in the future.

f) Test parameters

The process of computational validation is a continuing one, with varying goals, over a wide range of test conditions. For many applications at low supersonic speeds, the assumption of a perfect gas is a good one. The primary problem is one of physics and modeling, surface conditions, and flowfields. These parameters can be specified and experimentally determined. As the Mach number increases, however, real gas effects become important.

For applications where combustion takes place, for example in propulsion systems, the effect of chemistry becomes critical. For validation in these cases, a whole new set of parameters and required test conditions for the experiments and instrumentation are imposed. For such cases, not only is the usual fluid mechanics required for the initial definition, but also the total detailed chemical constitution of the flow.

3. *Concluding Remarks*

The present meeting is only one phase of a very complex computational validation process. In the following papers, the details of some new experiments will be described and the meeting will continue with comparison of many computations of these experiments. The experiments have been designed to cover certain physical phenomena. The author is sure that the process will continue with more complex experiments, more detailed instrumentation, and a continuing interaction with the computation, to design more critical experiments, requiring advanced instrumentation and measurements. This process is an important development in the establishment of the capability, accuracy, and confidence in CFD, for a range of design processes for many applications.

References

1. *Marvin, J. G.: Accuracy Requirements and Benchmark Experiments for CFD Validation. Symposium on Validation of Computational Fluid Dynamics, AGARD CP-437, 1988.*

2. *Goldhammer, M. and Rubbert, P.: CFD in Design: An Airframe Perspective. AIAA Paper 89-0092, January 1989.*

3. *Mehta, U. B.: Computational Requirements for Hypersonic Flight Performance Estimates (Invited). AIAA 89-1690, June 1989.*

HYPERSONIC STABILITY AND TRANSITION

Eli Reshotko
Case Western Reserve University
Cleveland, Ohio 44106, U.S.A.

ABSTRACT

The reliable computation of flow fields over hypersonic vehicles including skin friction and heat transfer information involves the determination of the location and extent of the transition region. The rational estimation of transition location is generally not a part of any contemporary flow field computational code. This paper discusses some of the factors that would have to be considered in the estimation of transition behavior for hypersonic flight.

This paper reviews the present state of our knowledge of the growing normal modes (the so-called second mode is the most dangerous at hypersonic speeds), and discusses the effects of surface cooling, pressure gradient and suction on their growth rates. Also discussed are effects of bluntness on the stability properties, and the aspects of environment and receptivity that are especially pertinent at hypersonic speeds.

The efficacy of the e^N method of transition estimation at supersonic and hypersonic speeds is reviewed. Comments are also made on the properties of hypersonic experimental facilities, and the care needed in doing effective experimental and computational study of this very delicate phenomenon.

1. INTRODUCTION

The evolution to turbulence in compressible boundary layers follows in the spirit of the "roadmap" of Figure 1 which illustrates the elements of the generally accepted view that laminar-turbulent transition is a consequence of the nonlinear response of the laminar boundary layer to forcing disturbances. These disturbances are part of the environment in which the laminar flow develops and could include freestream turbulence, freestream vorticity disturbances, entropy disturbances, radiated sound, surface roughness, surface vibrations, etc., or any combination of these.

In an environment in which initial disturbances are small (such as in flight), the transition Reynolds number depends on the nature

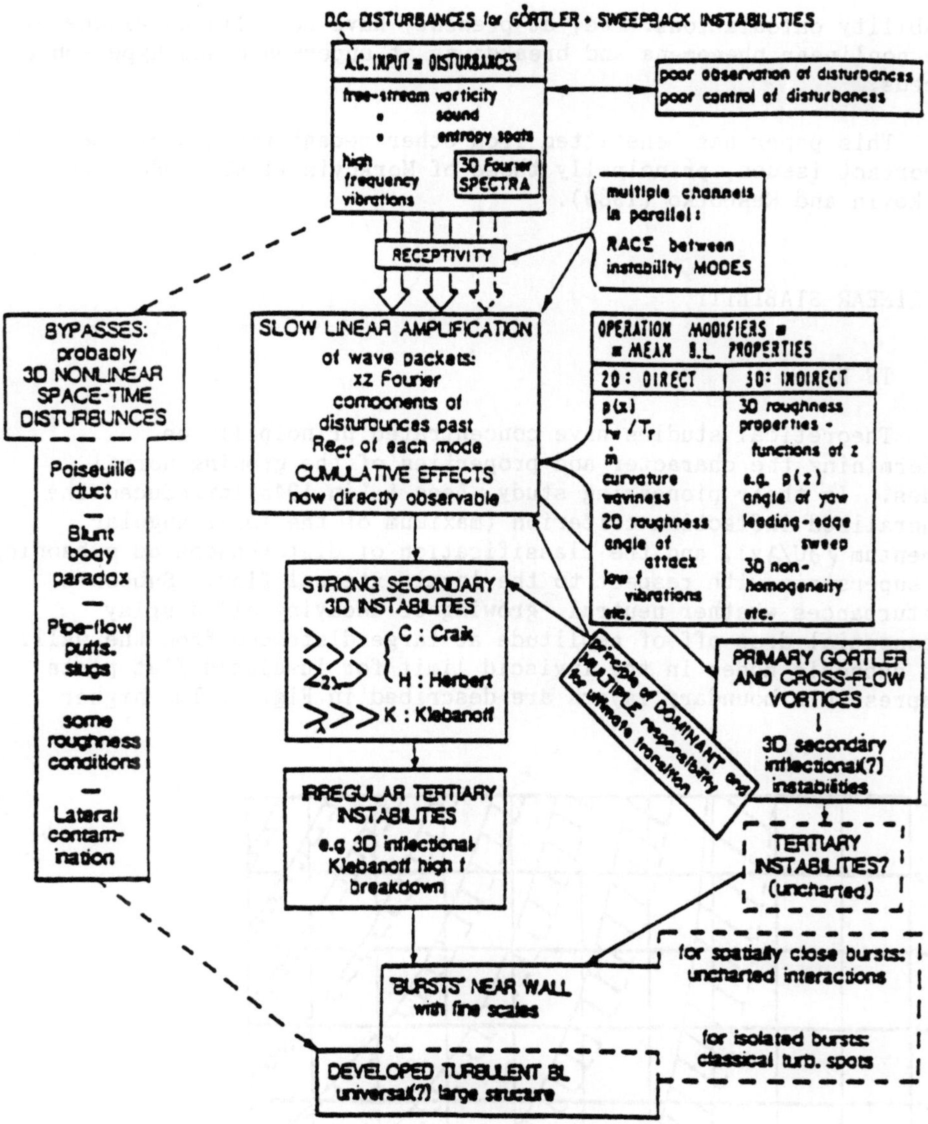

FIG. 1. Transition processes in wall layers

and spectrum of the disturbance environment, the signatures in the boundary layer of these disturbances and their excitation of the normal modes (receptivity), and importantly, the linear amplification of the growing normal modes. The wave interaction and nonlinear processes that follow serve to bring transition quickly to completion once they set in. It is expected that the factors affecting linear amplification are the primary factors that determine the magnitude of the transition Reynolds number. This is because the linear amplification step is the slowest of the successive multiple steps in the transition process. It is also the best understood. The most successful of the transition estimation techniques in use today are those based on the results of linear stability calculations. We, at present, have no solid knowledge of the nonlinear phenomena and breakdowns at supersonic and hypersonic speeds.

This paper has benefitted from other recent reviews of the important issues, principally those of Morkovin (1987, 1988) and Morkovin and Reshotko (1989).

2. LINEAR STABILITY

2.1 TS Waves

Theoretical studies have concentrated principally on determining the character and properties of the growing normal modes. In their pioneering study, Lees & Lin 1945 introduced the generalized inflection criterion (maximum of the local angular momentum $\rho\, \partial U/\partial y$), and the classification of disturbances as subsonic or supersonic with respect to the local external flow. Subsonic disturbances whether neutral, growing or decaying all display exponential drop-off of amplitude at large distances from the wall. The subsonic modes in the inviscid limit for insulated flat plate compressible boundary layers are described in Fig. 2. The higher

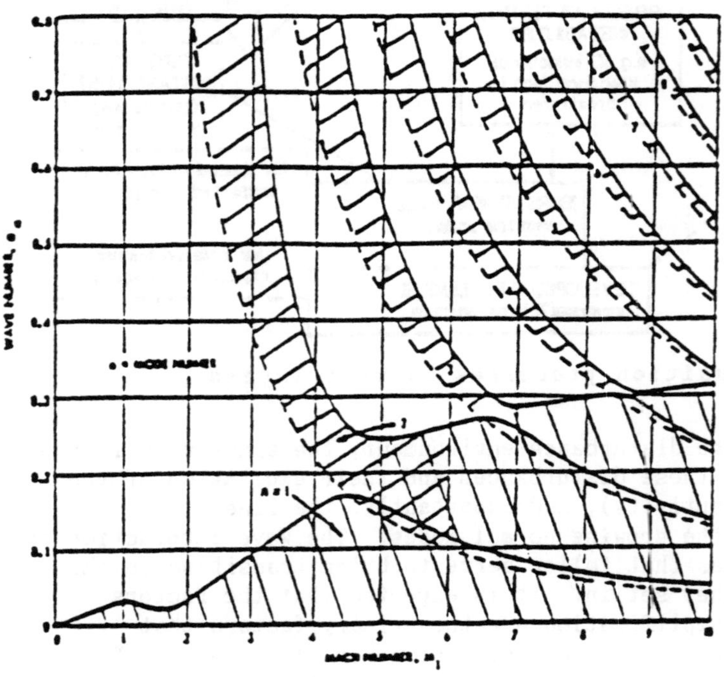

FIG. 2. Wave numbers $2\pi L^*/\lambda^*$ corresponding to inviscid instabilities in insulated flat-plate boundary layers.

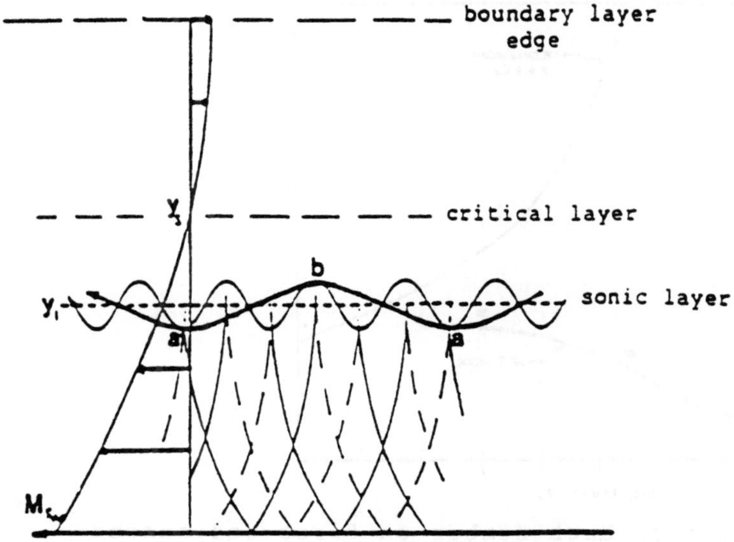

wall recedes "supersonically" from wave

FIG. 3. Mack's neutral trapped acoustic mode

subsonic modes discovered by Mack (1965) are trapped acoustic modes (see Fig. 3) that occur when additionally the wall is supersonic relative to the phase velocity ($c_r/a_w > 1$). For insulated surfaces, these higher modes appear for $M > 2.2$; however it is not until the Mach number is of the order of 4 or greater that the second mode is at low enough frequency to have experimental consequences (Fig. 4). Once the second mode sets in, it becomes the dominant instability since its growth rate tends to exceed that of the first mode. For insulated surfaces, this occurs for $M > 4$ (Fig. 5). For cooled surfaces, second mode dominance can occur at even lower Mach number. In contrast to the first or TS mode, the higher subsonic modes are destabilized by cooling (Mack, 1969). Preliminary indications are that second mode subsonic disturbances can be stabilized by favorable pressure gradient or by suction (Malik 1988, Malik 1989, Mack 1989).

Supersonic disturbances ($c_r < 1 - 1/M_e$) were long considered irrelevant since supersonic neutral disturbances have finite amplitudes to infinity in y. Mack (1969) has shown however that amplifying outgoing supersonic disturbances do decay exponentially for large y and are therefore acceptable normal modes (associated physical reasoning remains to be developed). Quantitatively, Mack (1969) found second mode supersonic disturbances at Mach 5.8 for highly cooled walls ($T_w/T_{aw} < 0.15$) in the inviscid limit. Their growth rates however were found to be much lower than those of the subsonic disturbances and their wavenumbers (frequencies) noticably larger. More recently, Mack (1989) has calculated cases of highly cooled boundary layers at $M = 3$ and $M = 4.5$, again in the inviscid limit, and found second mode supersonic disturbances that are destabilized by suction. Although the growth rates of these latter

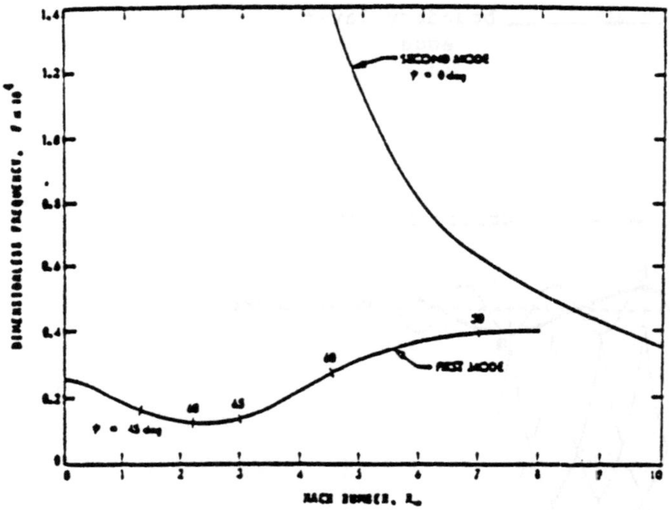

FIG. 4. Effect of Mach number on first-and second-mode most unstable frequencies at Re=1500. Insulated wall; wind-tunnel temperatures (Mack 1969)

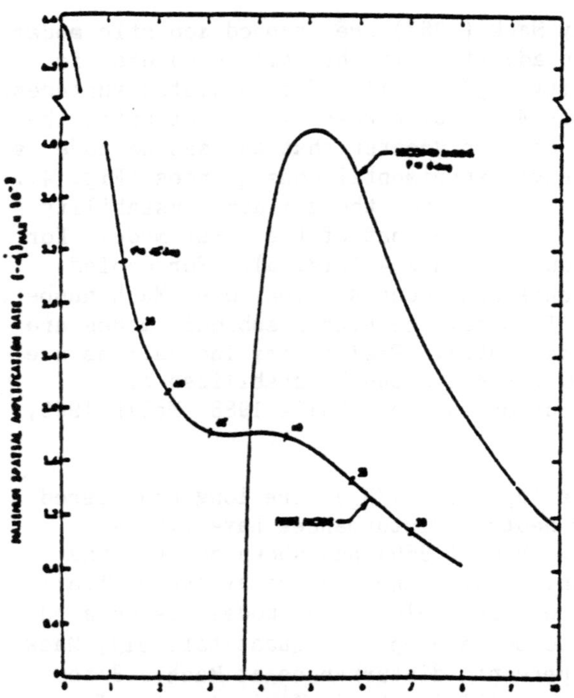

FIG. 5. Effect of Mach number on spatial amplification rate of most unstable first-AND second-mode disturbances at Re=1500. Insulated wall; wind tunnel temperatures (Mack 1969)

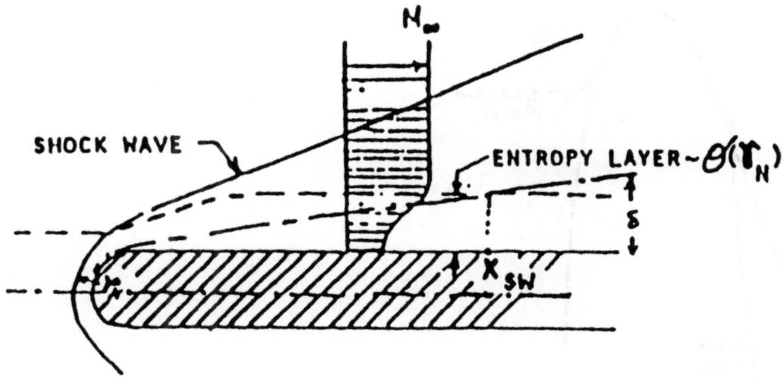

FIG. 6. Sketch of the flow field of a blunted plate in a supersonic stream

disturbances approach the levels associated with subsonic disturbances, the disturbance wavenumbers and corresponding frequencies are very large. The significance of these disturbances to the transition process is likely to be negligible since Mack (1990) further indicates that they are strongly damped at finite Reynolds number.

2.2 Effects of Bluntness

The effects of leading edge or nose bluntness on TS stability have only recently been analyzed beyond the work of Reshotko and Khan (1980). These authors showed that for a blunted flat plate, upstream of swallowing, both the boundary layer and the thicker shock layer can be separately unstable in a generalized inflectional sense (Fig. 6). After the BL swallows the entropy layer, the boundary layer profiles asymptote to those for a sharp leading edge and their stability characteristics follow suit. Within the swallowing region, the stability characteristics are also affected by the fact that the shock layer flow at the edge of the BL is non-uniform in y. This seems to be stabilizing for the first mode but destabilizing for the second mode (Fig. 7). Equivalent theoretical studies for blunted cones have been hampered by the unavailability of reliable laminar flowfield calculations particularly over the first 20 or so nose radii. Recent computational results by Malik (1988) for the blunted cone of the experiment of Stetson et al (1984) ($M_\infty = 8$, $\theta_c = 7^o$) qualitatively resemble experimental features including presence of the second-mode frequency band (Fig. 8). But they significantly underestimate $Re_{min,crit}$ for both the sharp and blunted cones. For the sharp cone, this may be due to very small bluntness of the experimental cone. For the blunted cone, the discrepancy may stem from an inadequate calculation of the basic flow in the vicinity of the nose. An elaboration of this work is described in the paper by Malik, Spall & Chang (1990) who point out further that their computations did not detect the instability in the shock layer as observed by Stetson et al (1984). No theory

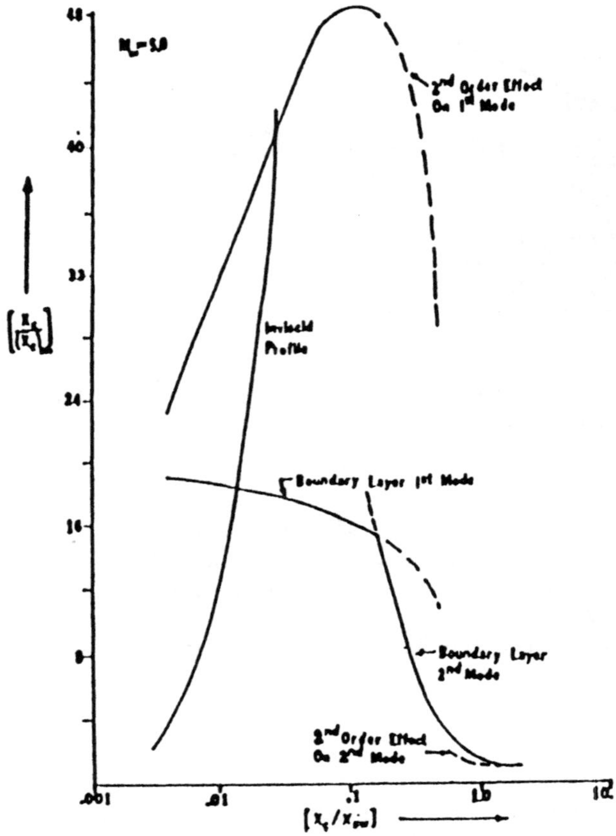

FIG. 7. **Effect of bluntness on location of minimum critical point at M = 5.0 (Reshotko & Khan 1980)**

offers explanations for the observed growth in a frequency band above the second mode but below the third. Stetson speculatively identifies this feature as a nonlinearity. This unsatisfactory state reflects the paucity of consistent matched theoretical and experimental information on stability characteristics at supersonic and hypersonic speeds.

2.3 Finite Shock-Layer Effects

All the TS stability calculations to date except those of Reshotko & Khan (1980) assume the external flow to be uniform "to infinity," and therefore have normal mode eigenfunctions that decay exponentially to infinity in y. For hypersonic flows where shock layers are thin, the outer boundary conditions for the disturbance flow may have to be applied at the shock wave. A first attempt at formulating such boundary conditions is by Cowley & Hall (1988). They however assume that the undisturbed flow is uniform in y below the shock wave which may not always be the case.

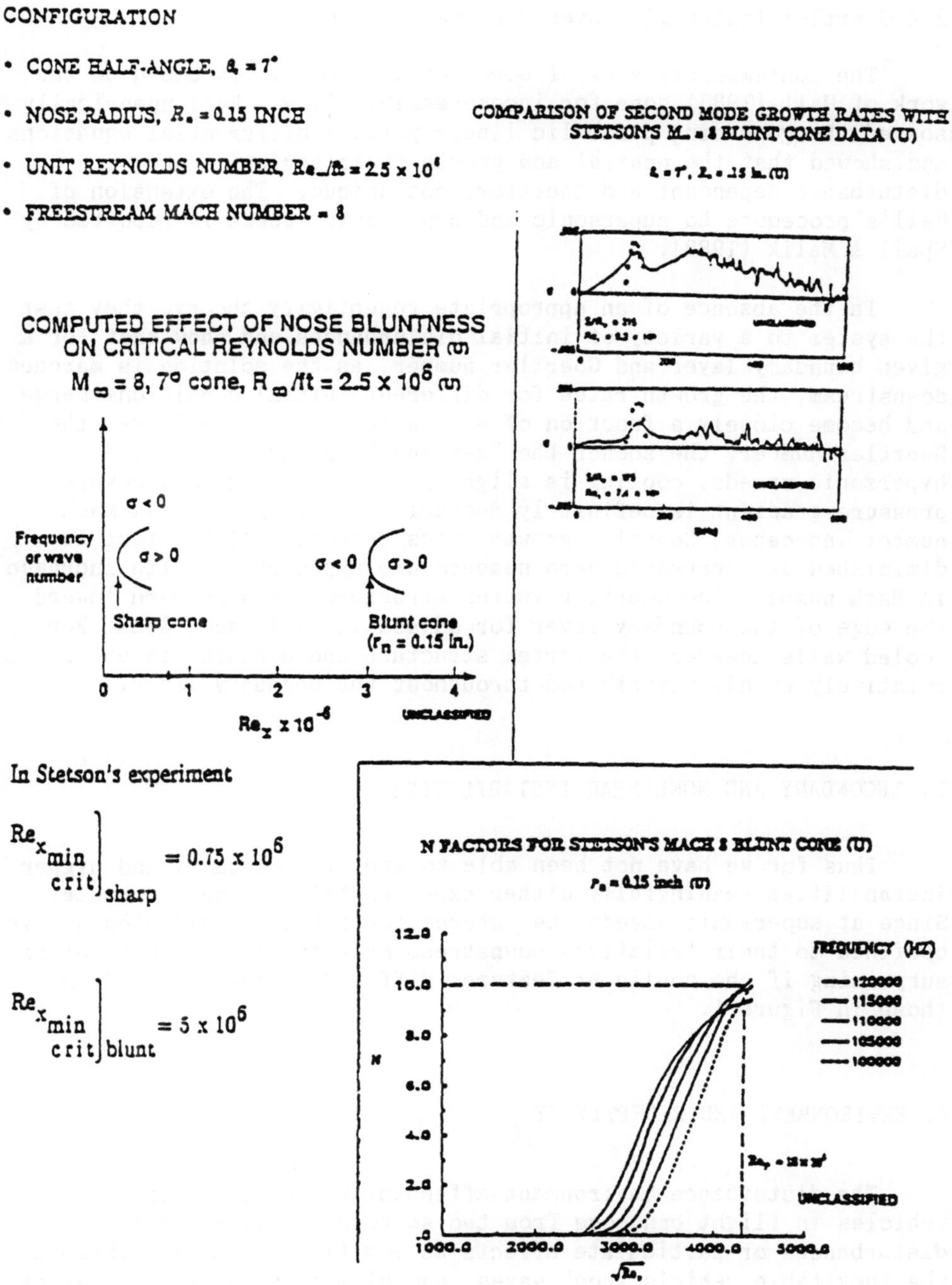

FIG. 8. Calculations of effects of bluntness relating to Stetson's blunt cone experiment (Malik 1988)

2.4 Goertler Instability over Concave Surfaces

The contemporary view of Goertler instability is based on the work of Hall (1983) done for incompressible flow. Hall numerically solved the governing parabolic linear partial differential equations and showed that the neutral and growth characteristics are disturbance dependant and therefore not unique. The extension of Hall's procedure to supersonic and hypersonic speeds is reported by Spall & Malik (1989).

In the absence of an appropriate receptivity theory, they test the system to a variety of initial disturbances and show that for a given boundary layer and Goertler number, as the solution is marched downstream, the growth rates for different initial conditions merge and become closely a function of wavenumber only. The larger the Goertler number, the sooner the "merging." At supersonic and hypersonic speeds, cooling is slightly destabilizing, and adverse pressure gradient is definitely destabilizing (Fig. 9). As Mach number increases, Goertler growth rates decrease. The effect is diminished as hypersonic Mach numbers are approached. With increase in Mach number, the Goertler vortex structure moves outward toward the edge of the boundary layer for adiabatic wall conditions. For cooled walls however, the vortex structure and disturbance energy is relatively evenly distributed throughout the boundary layer.

3. SECONDARY AND NONLINEAR INSTABILITIES

Thus far we have not been able to explore secondary and higher instabilities meaningfully either experimentally or on a computer. Since at supersonic speeds the interaction between fluid elements is confined to their (relative) downstream Mach cones, it would not be surprising if the nonlinear features differed substantially from those in Figure 1.

4. ENVIRONMENT AND RECEPTIVITY

The disturbance environment affecting boundary layers on vehicles in flight can come from two sources: a) atmospheric disturbances or particulate effects as modified by passage through the inevitable vehicle shock waves, and b) mechanical and acoustical disturbances coming from other parts of the vehicle. Bushnell (1989) supplies background, references and speculations on triggering mechanisms associated with such disturbances. His information on particulates in the atmosphere and their role, and on the possibility of transition promotion by sustained high-frequency electrostatic discharges in flight is thought provoking.

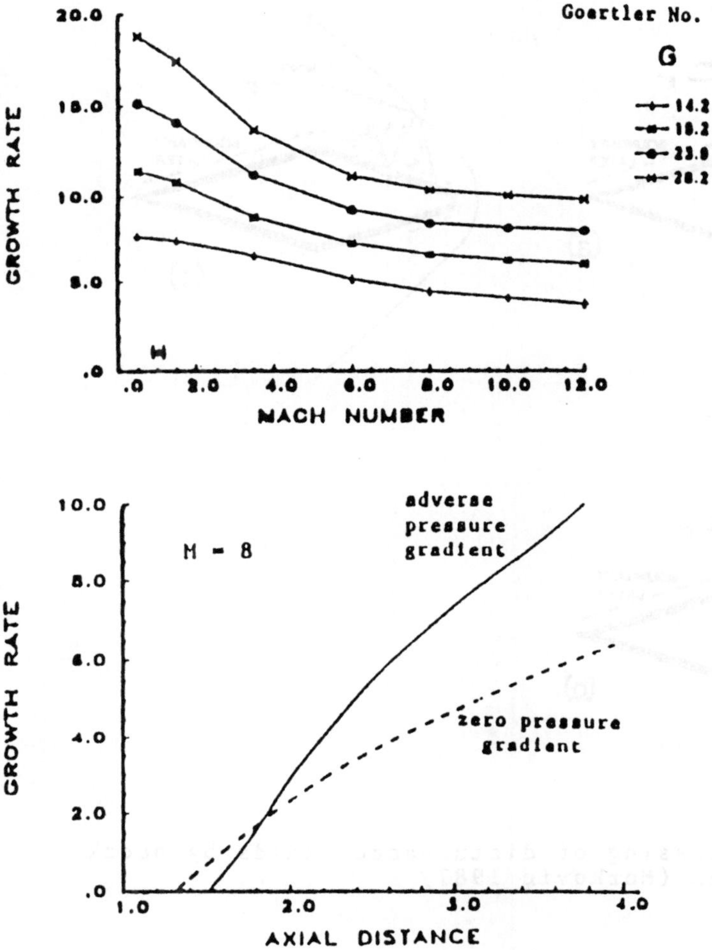

FIG. 9. **Effects of pressure gradient and Mach number on growth rates of Goertler instability (Spall & Malik 1989)**

Receptivity phenomena for supersonic and hypersonic boundary layers have yet to receive serious attention. Whatever the receptivity mechanisms, the receptivity considerations are altered considerably by the presence of attached or bow shock waves and the fact that while vorticity and entropy disturbances are convected along streamlines, acoustic disturbances propagate along (relative) Mach lines (Fig. 10). Any free-stream disturbance field is altered in passage through the shock wave. When entropy or vorticity disturbances on streamlines remote from the boundary layer pass through the shock wave, they are modified and are partially converted to pressure waves that impinge on the boundary layer along (relative) Mach lines. Sound waves change amplitude and refract through shock waves and give rise to vorticity and entropy disturbances. Furthermore, irregularities on the vehicle surface generate pressure waves which are modified by the shock and

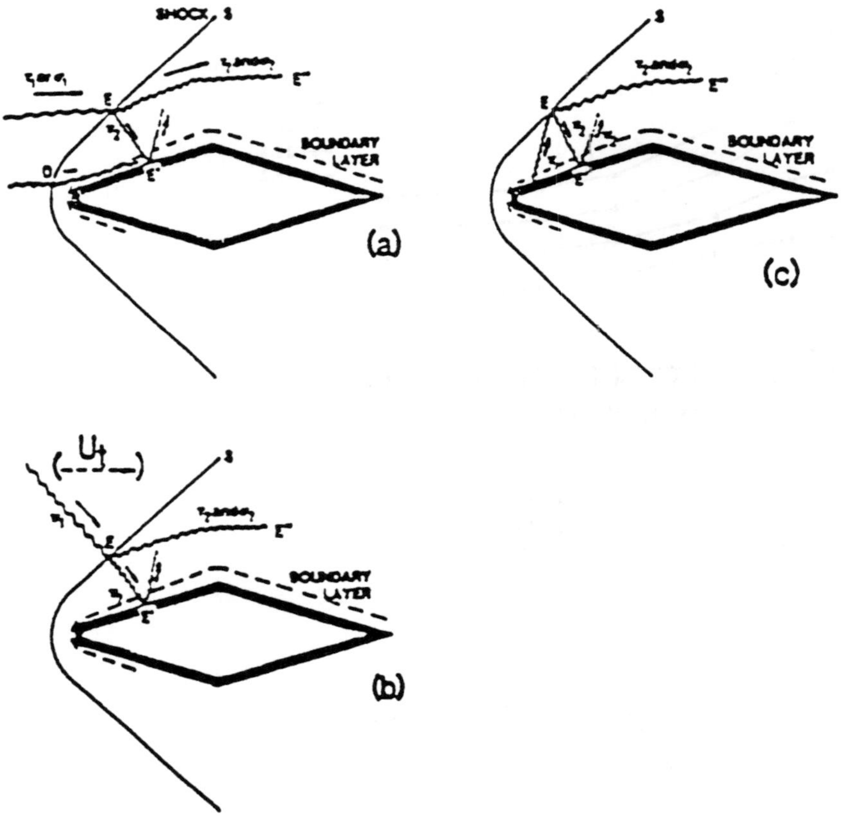

FIG. 10. Processing of disturbance fields by shock waves (Morkovin 1987)

reflected back toward the body. Once the disturbances are internalized in the boundary layer, the problem of matching wavenumbers and frequencies of amplified normal modes remains conceptually similar to that at low speeds.

5. TRANSITION ESTIMATION

Current exp N techniques for estimating transition location are based on linear stability calculations of the slowly growing instabilities (TS, crossflow, Goertler) following the suggestions of Smith & Gamberoni (1956 and van Ingen (1956). The essential features of the underlying "correlation" between the experimental transition location x_{tr} and the inferred growth factor for transition, N_{tr}, are outlined in Note A-3 of Morkovin & Reshotko (1989), together with a list of uncertainties inherent in the process. The strength of the exp N method lies in its flexibility in design, where a number of alternative vehicles are evaluated for trade-off analyses. The assumption of the method make it possible to compare their transition behavior (and hence drag, local heating, etc.) without any knowledge of the environmental disturbances. The effect of control strategies can be incorporated in the comparative analysis.

Bushnell claims success using the exp N method for flat plates and cones when comparing with data from flight experiments and with data from the NASA-Langley M = 3.5 pilot quiet tunnel. Calculations by Malik (1988, 1989) using an N_{tr} of 9 appear in reasonable agreement with a number of data sets. Perhaps more significantly, the calculations offer a way of distinguishing between those regions of parameter space where transition results from first mode instability and those where transition results from second mode instability (Fig. 11).

Transition estimation procedures for three dimensional boundary layers such as found on swept wings and yawed bodies are far from settled. On practical configurations one may have to contend with TS, crossflow and even Goertler instabilities simultaneously. Since the boundary layer properties change in both x and z, the wave fronts change directions and the x-z paths of growing wave packets are not obvious.

6. EXPERIMENTAL STUDIES OF STABILITY AND TRANSITION

It has long been known that transition behavior in conventional supersonic wind tunnels above M = 2 can be clearly ascribed to the noise radiated onto the model from the turbulent boundary layers on the tunnel walls (Pate & Schueler, 1969). This was confirmed again within the last decade by the important flight experiments of Dougherty and Fisher (1980). Their transition cone model had

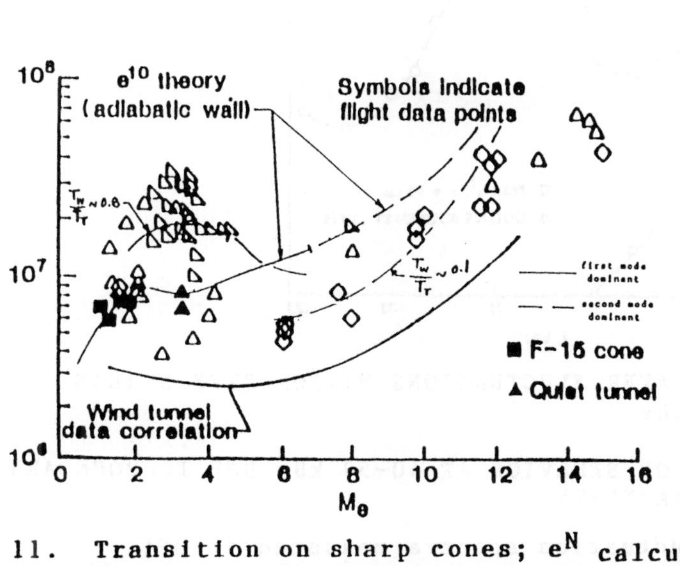

FIG. 11. Transition on sharp cones; e^N calculations compared with flight data (Malik 1988)

already been run in many transonic and supersonic wind tunnels throughout the world. In wind tunnels for M > 2 they consistently obtained the familiar transition "bucket" with a minimum of Re_{tr} between Mach numbers 4 and 5. However, in flight experiments with the cone model mounted ahead of the fuselage of an F-15 aircraft, the transition Reynolds number corrected to adiabatic wall conditions showed consistent increase with Mach number over the entire flight speed range, whereas the values for Re_{tr} obtained in conventional wind tunnels started falling at M > 1.6. This further accentuated the need for further flight experiments and for studies in "quiet" supersonic tunnels. The character, limitations and accomplishments of the NASA-Langley pilot quiet tunnel are highlighted in the recent paper of Beckwith & Chen (1989).

How does one define a "quiet" tunnel? Seemingly it should be one where the frequency bands of unstable modes are not excited by acoustic radiation from the turbulent boundary layers on the tunnel walls. "Conventional" quiet tunnels do this by maintaining laminar boundary layers on the tunnel walls. At M > 8, even conventional tunnels with turbulent sidewalls display an essentially quiet behavior as the instability bands tend to be separated from the tunnel noise spectrum (Fig. 12). These possibilities merit further investigation.

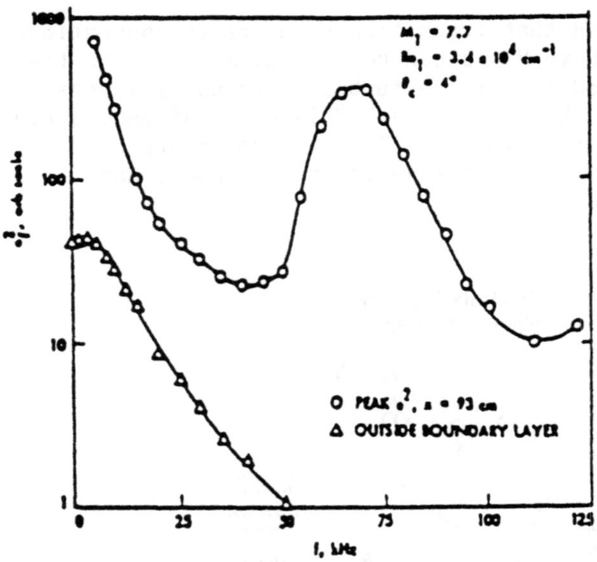

- BOUNDARY LAYER FLUCTUATIONS MIRROR FREE STREAM BELOW 35 kHz

- DEPARTURE OF BEHAVIOR AT 40-50 kHz DUE TO BOUNDARY LAYER INSTABILITY

FIG. 12. Fluctuation spectra measured in JPL 21" tunnel at M_{∞} = 8.5 (Kendall 1975)

Conventional supersonic tunnels may still be used for TS, swept attachment-line and cross-flow instability studies, provided that it is ascertained that the results are not affected by tunnel noise.

The standards for research quality experiments on stability and transition are the "guidelines" as formulated by the U.S. Boundary Layer Transition Study Group (Reshotko 1976):

1. Any effects specifically and only associated with test facility characteristics must be identified and if possible avoided.

2. Attention must be given to disturbances introduced by the model surface, model material and internal structure. Experimental studies should include documentation of these various factors.

3. Details of coupling of disturbamces of various kinds to the boundary layer must be understood theoretically and experimentally, so that the sensitivity of the transition process to the flight environment might be determined.

4. Whenever possible, tests should involve more than one facility. Tests should have ranges of overlapping parameters, and whenever possible, experiments should have redundancy in transition measurements.

Morkovin has suggested strongly that Guideline Number Four be applied also to computational studies.

The above guidelines apply as well to flight experiments. Their implementation for flight experiments requires special attention to a number of factors:

1. The measurement of disturbance environment must be incorporated into the model design and in fact must be part of the model.

2. Attention has to be given to the maintenance and monitoring of test conditions such as Mach number, Reynolds number, angle of attack, yaw angle and surface temperature for the duration of the measurement period.

3. Attention has to be given to the maintenance and monitoring of model surface conditions for each flight. This includes protection of the model surface before launch, and recovery of the vehicle for inspection and reconditioning of the surface prior to the succeeding flight.

4. Because stability phenomena at supersonic and hypersonic speeds occur at frequencies of hundreds of kHz and even to MHz levels, there is a need for very high data sampling rates, especially when monitoring multiple channels. This

poses special problems in data acquisition and data reduction. Reliable digital telemetering of data from the vehicle may also be necessary in order to minimize weight and volume of the data acquisition equipment.

7. DIRECT COMPUTATIONAL SIMULATION OF TRANSITION

One can certainly envision (or dream about) an eventual capability of computing the full flowfield about a vehicle or vehicle component including the transition process. Since transition is the result of unsteady phenomena as discussed in this paper, the computation through transition requires solution of compressible three-dimensional unsteady, time-accurate Navier-Stokes equations in a domain large enough to contain the complete phenomenon, and with a grid that is fine enough over the extended boundary layer to capture the detailed linear eigenfunctions and the subsequent nonlinearities. In addition, initial and boundary conditions must be properly specified and enforced. The initial and upstream boundary conditions should represent properly the physical disturbance environment to which the flow is subjected. Truncation error and round-off introduce disturbances whose spectrum is non-physical. The downstream boundary condition should be non-reflecting in order that spurious feedback phenomena be absent. (See Herbert 1989 for critical discussion of relevant issues).

At present, these requirements are marginally attainable only for small portions of simple flows. The calculation of a full vehicle is still a number computer generations ahead of us. Parallelization of the codes and running them on multiprocessors can result in significant reductions in running time (Erlebacher et al 1990). But until the overall capability is available, the estimation of transition locations will probably have to depend on linear stability properties of the calculated laminar flows and their synthesis into an exp N procedure.

8. CLOSING COMMENT

The tendency in flowfield calculation at supersonic and hypersonic speeds has been to use elaborate three-dimensional codes over appropriately shaped and detailed grids to calculate the steady laminar and Reynolds-averaged turbulent flowfields for the configurations of interest. Many of these codes incorporate a transition "switch" using algebraic correlations that have little physical basis and much data scatter. If the steady flow is already so complicated, why should one expect transition, which is a nonlinear consequence of the unsteady flow, to be governed by simple formulas?

Clearly the rational treatment of transition in a computational code must involve consideration of the physics of the process as discussed in this paper. To do less is to invite serious misjudgments in vehicle design.

This work was supported by the NASA-Lewis Research Center under Grant NCC3-124.

REFERENCES

Beckwith,I.E., Chen,F.J. 1989. Paper 4.1, Third IUTAM Symposium on Laminar-Turbulent Transition, Toulouse, France

Bushnell,D.M. 1989. Notes on initial disturbance fields for the transition problem. NASA, NASP Tech. Memo. 1051

Cowley,S., Hall,P. 1988. On the instability of hypersonic flow past a wedge. ICASE Report No. 88-72 (NASA Contractor Rept. 181760)

Dougherty,N.S.Jr., Fisher,D.F. 1980. Boundary layer transition on a 10-degree cone. AIAA Paper 80-0154

Erlebacher,G., Bokhari,S.H., Hussaini,M.Y. 1990. Parallelization of a Three-Dimensional Compressible Transition Code. AIAA Jour. v.28, no.1, pp. 83-90

Hall,P. 1983. The linear development of Goertler vortices in growing boundary layers. J. Fluid Mech. v.130, pp. 41-58

Herbert,T. 1989. Exploring transition by computer. to appear in J. of Appl. Num. Math.

Kendall,J.M. 1975. Wind tunnel experiments relating to supersonic and hypersonic boundary-layer transition. AIAA J., v.13, no.3, pp. 290-299

Lees,L., Lin,C.C., 1946. Investigation of the stability of the laminar boundary layer in a compressible fluid. NACA TN 1115

Mack,L.M. 1965. Stability of the compressible laminar boundary layer according to a direct numerical solution. AGARDograph 97, pt.1, pp. 329-362

Mack,L.M. 1969. Boundary Layer Stability Theory. JPL Rept. 900-277, Rev. A

Mack,L.M. 1989. On the Inviscid Acoustic-Mode Instability of Supersonic Shear Flows. To appear in Journal of Theoretical and Computational Fluid Mechanics

Mack,L.M. 1990. Private communication

Malik,M.R. 1988. Stability theory for NASP transition prediction. Fifth National Aero-Space Plane Symposium, Paper no. 57

Malik,M.R. 1989. Prediction and Control of Transition in Supersonic and Hypersonic Boundary Layers. AIAA Jour. v.27, no.11, pp.1487-1493

Malik,M.R., Spall,R.E., Chang,C.-L. 1990. Effect of Nose Bluntness on Boundary Layer Stability and Transition. AIAA Paper 90-0112

Morkovin,M.V. 1987. Transition at Hypersonic Speeds. ICASE Interim Report 1 (NASA Contractor Report 178315)

Morkovin,M.V. 1988. Recent insights into instability and transition to turbulence in open-flow systems. AIAA Paper 88-3675. Also ICASE Report No. 88-44, (NASA Contractor Report 181693)

Morkovin,M.V., Reshotko,E. 1989. Dialogue on Progress and Issues in Stability and Transition Research. Proc. Third IUTAM Symposium on Laminar-Turbulent Transition, Toulouse, France

Pate,S., Schueler,C. 1969. Radiated aerodynamic noise effects on boundary-layer transition in supersonic and hypersonic wind tunnels, AIAA J., v.7, pp. 450-457

Reshotko,E. 1976. Boundary layer stability and transition. Ann. Rev. of Fluid Mech. v. 8 pp. 311-350

Reshotko,E., Khan,M.M.S., 1980. Stability of the laminar boundary layer on a blunted plate in supersonic flow. in Laminar-Turbulent Transition, IUTAM Symposium Stuttgart 1979, Springer-Verlag

Smith,A.M.O., Gamberoni,N. 1956. Transition, pressure gradient and stability theory, Douglas Aircraft Co. Report No. ES 26388

Spall,R.E., Malik,M.R. 1989. Goertler vortices in supersonic and hypersonic boundary layers. Phys. Fluids A 1 (11) pp. 1822-1835

Stetson,K.F., Thompson,E.R., Donaldson,J.C., Siler,L.G. 1984. Laminar boundary layer stability experiments on a cone at Mach 8, part 2: blunt cone. AIAA Paper 84-0006

Van Ingen,J.L. 1956. A suggested semi-empirical method for the calculation of the boundary-layer transition region. Dept of Aero. Eng., Delft, Reports VTH-71 and VTH-74.

SCATTERING KERNELS FOR GAS-SURFACE INTERACTION

Carlo Cercignani
Dipartimento di Matematica
Politecnico di Milano
Piazza Leonardo da Vinci,32
20133 Milano (Italy)

Abstract. The problem of the interaction between a gas molecule and the surface layers is attacked from a semiphenomenological point of view. After extracting from physics all the general information which is available, theoretical models embodying this information, sufficiently simple and flexible to cope with different physical situations, are discussed. Other approaches and comparisons with experimental data are also reviewed.

1. Introduction.

When a gas flows past a solid body (or in a region bounded by one or more solid bodies), the Boltzmann equation must be accompanied by boundary conditions, which describe the interaction of the gas molecules with the solid walls. It is to this interaction that one can trace the origin of the drag and lift exerted by the gas on the body and the heat transfer between the gas and the solid boundary.

The difficulties of a theoretical investigation are due, mainly, to our lack of knowledge of the structure of surface layers of solid bodies and hence of the effective interaction potential of the gas molecules with the wall. When a molecule impinges upon a surface, it is adsorbed and may form chemical bonds, dissociate, become ionized or displace surface molecules. Its interaction with the solid surface depends on the surface finish, the cleanliness of the surface, its temperature, etc. It may also vary with time because of outgassing from the surface. Preliminary heating of a surface also promotes purification of the surface through emission of adsorbed molecules. In general, adsorbed layers may be present; in this case, the interaction of a given molecule with the surface may also depend on the distribution of molecules impinging on a surface element.

The first observations of the phenomenon of slip of a gas on a solid surface are due to Kundt and Warburg[1]. Maxwell[2] had already suggested that slip would be a consequence of kinetic theory and computed the amount of slip for a particular model of the gas-surface interaction, still very popular and known under his name. Smoluchowski[3] described temperature jump in a similar way. A more systematic effort started with Knudsen[4]. Full-scale research, however, only began in the late 1950's, under the impetus of space flight and thanks to the developments in high vacuum technology.

The possible events, that a gas molecule hitting a solid (or liquid) surface may experience, have been briefly described above. A more systematic classification, following I. Kuščer[5] but adding reactive interactions to his list, may read as follows:

1. Scattering.
   1.1 Elastic scattering (zero-phonon scattering):
      a) specular reflection;
      b) diffraction into discrete peaks;
      c) diffuse elastic scattering.
   1.2 Inelastic scattering:
      a) one-phonon scattering;
      b) multi-phonon scattering.
   1.3 Sputtering and other high energy effects.
2. Adsorption.
   2.1 Adsorption at specific sites:
      a) temporary;
      b) permanent.
   2.2 Mobile adsorption (surface diffusion).
   2.3 Diffusion into the condensed phase.
3. Condensation (in case of identical composition of gas and condensed phase).
4. Reactive interaction.
   4.1 Chemical reactions.
   4.2 Ionization.

Elastic scattering is strong only if the surface behaves as it were almost rigid; this may occur if both the energy of the gas molecule and the thermal energy of the wall are comparable to or below $k_B T_D$ (where k_B is the Boltzmann constant and T_D the Debye temperature). In this case specular reflection can occur only if the equipotential surfaces are approximately flat on the molecular scale. Thermal motions in the solid reduce elastic scattering, till, if the molecule's kinetic energy is several times in excess of $k_B T_D$, incident molecules become capable of ejecting atoms or whole chunks of the solid ("sputtering").

Gas molecules with strong chemical affinity to the solid become adsorbed, *i. e.* attached to the surface. Even in the absence of such affinity, attachment can occur at sufficiently low temperatures, because of van der Waals forces. Although the difference is mainly in the magnitude of the binding energy, one usually distinguishes the two cases by means of the names chemisorption and physisorption. Any adsorption is, in principle, temporary[6], but for strong binding and low surface temperatures the sitting time may well exceed the duration of any experiment or even the age of the Universe; in this case one talks about permanent adsorption. Heating strongly accelerates the desorption process. Strongly adsorbed molecules are usually localized at particular adsorption sites, but even a face of an ideal crystal offers several kinds of such sites. Polycrystals and amorphous solids are expected to be far richer in this respect. The behavior of the adsorbed substance very much depends on the fraction of adsorption sites actually occupied. A complete layer looks like a part of the crystal and has a similar if not the same symmetry. Above it a second layer can sometimes be adsorbed.

The exchange of matter in condensation and evaporation is analogous to adsorption and desorption, except for the identical composition of both phases. This has the important consequence that the surface remains unchanged.

Chemical reactions may occur at the wall; *e. g.*, the adsorbate can dissociate during adsorption, or, if the gaseous phase is a mixture, the wall can catalyze reactions that would not occur in the gas. Analogously one may have ionization of the adsorbate.

2. The scattering kernel.

In general, a molecule striking a surface with a velocity $\mathbf{v}'$ reemerges from it with a velocity $\mathbf{v}$ which is strictly determined only if the path of the molecule within the wall can be computed exactly. This computation is very hard, because it depends upon a great number of details, such as the locations and velocities of all the molecules of the wall and an accurate knowledge of the interaction potential (see Sects. 4 and 6). In fact a sufficiently accurate calculation should be able to predict all the phenomena that we have have briefly described in the previous section.

Hence it is more convenient to think in terms of a probability density $R(\mathbf{v}'\to\mathbf{v};\mathbf{x},t;\tau)$ that a molecule striking the surface with velocity between $\mathbf{v}'$ and $\mathbf{v}'+d\mathbf{v}'$ at the point $\mathbf{x}$ and time t will re-emerge at practically the same point with velocity between $\mathbf{v}$ and $\mathbf{v}+d\mathbf{v}$ after a time interval τ (adsorption or sitting time). If R is known, then we can easily write down the boundary condition for the distribution function $f(\mathbf{x},\mathbf{v},t)$. To simplify the discussion, the gas will be presently assumed to be monatomic and the surface at rest.

The mass (or number, depending on normalization) of molecules emerging with velocity between $\mathbf{v}$ and $\mathbf{v}+d\mathbf{v}$ from a surface element dA about $\mathbf{x}$ in the time interval between t and t+dt is

$$d^{*}\mathcal{M} = f(\mathbf{x},\mathbf{v},t)|\mathbf{v}\cdot\mathbf{n}|dtdAd\mathbf{v} \qquad (\mathbf{x}\in\partial\Omega,\ \mathbf{v}\cdot\mathbf{n}>0) \qquad (2.1)$$

where $\mathbf{n}$ is the unit vector normal to the surface $\partial\Omega$ at $\mathbf{x}$ and directed from the wall into the gas. Analogously, the probability that a molecule impinges upon the element dA with velocity between $\mathbf{v}'$ and $\mathbf{v}'+d\mathbf{v}'$ in the time interval between $t-\tau$ and $t-\tau+dt$ ($\tau>0$) is

$$d^{*}\mathcal{M}' = f(\mathbf{x},\mathbf{v}',t-\tau)|\mathbf{v}'\cdot\mathbf{n}|dtdAd\mathbf{v}' \qquad (\mathbf{x}\in\partial\Omega,\ \mathbf{v}'\cdot\mathbf{n}<0) \qquad (2.2)$$

If we multiply $d^{*}\mathcal{M}'$ by the probability of a scattering event from velocity $\mathbf{v}'$ to a velocity between $\mathbf{v}$ and $\mathbf{v}+d\mathbf{v}$ with an adsorption time between τ and $\tau+d\tau$ (*i. e.* $R(\mathbf{v}'\to\mathbf{v};\mathbf{x},t;\tau)d\mathbf{v}d\tau$) and integrate over all the possible values of $\mathbf{v}'$ and τ, we must obtain $d^{*}\mathcal{M}$ (here we assume that each molecule re-emerges from the surface element into which it entered, which is not so realistic when τ is large):

$$d^{*}\mathcal{M} = d\mathbf{v}\int_0^\infty d\tau\int_{\mathbf{v}'\cdot\mathbf{n}<0} R(\mathbf{v}'\to\mathbf{v};\mathbf{x},t;\tau)d^{*}\mathcal{M}' \qquad (\mathbf{x}\in\partial\Omega,\ \mathbf{v}\cdot\mathbf{n}>0) \qquad (2.3)$$

Equating the expressions in Eqs. (2.1) and (2.3) and cancelling the common factor dAd$\mathbf{v}$dt, we obtain

$$f(\mathbf{x},\mathbf{v},t)|\mathbf{v}\cdot\mathbf{n}| = \int_0^\infty d\tau\int_{\mathbf{v}'\cdot\mathbf{n}<0} R(\mathbf{v}'\to\mathbf{v};\mathbf{x},t;\tau)f(x,\mathbf{v}',t-\tau)|\mathbf{v}'\cdot\mathbf{n}|d\mathbf{v}'$$

$$(\mathbf{x}\in\partial\Omega,\ \mathbf{v}\cdot\mathbf{n}>0) \qquad (2.4)$$

If $\bar{\tau}$ is an average adsorption time, $\bar{v}$ the average normal velocity with which the gas molecules impinge upon the surface and n the number density of the gas, $n\bar{v}dA$ molecules will impinge, per unit time, on a surface element of area dA and stay there an average time $\bar{\tau}$; if σ_0 is an effective range of the gas-surface interaction, each molecule will occupy an area of the order of $\sigma_0^{\,2}$ and the total area occupied by adsorbed molecules will be $n\overline{v\tau}\sigma_0^{\,2}dA$, *i. e.* a fraction $n\overline{v\tau}\sigma_0^{\,2}$ of the surface will be occupied. This is, of course, just a rough order-of-magnitude argument, since the molecules may penetrate somewhat into the solid and not necessarily remain at its surface.

If $n\sigma_0^{\,2}\overline{v\tau}$ is not close to zero, the nature of the interaction of each incident molecule depends on the total number and energy of the incident molecules. Under conditions of extremely low density (as, for example, in the case of an orbiting satellite) $n\sigma_0^{\,2}\overline{v\tau}<<1$ and each incident molecule interacts with the surface independently of the others. The same independence may show up in the other limiting case $n\sigma_0^{\,2}\overline{v\tau}\cong 1$ (for example, in chemisorption when $\bar{\tau}$ may be very large, or in a dense gas when n can be extremely large); in this case the impinging molecule interacts directly with the adsorbed layer rather than with the atoms of the solid (and the effective adsorption time for that interaction can be much shorter than $\bar{\tau}$).

Whenever the effective adsorption time $\bar{\tau}_e$ and the number density are such that $n\sigma_0^{\,2}\overline{v\tau}_e<<1$, scattering events can safely be considered instantaneous and statistically independent and we can assume that the kernel $R(\mathbf{v}'\to\mathbf{v};\mathbf{x},t;\tau)$ does not depend on the distribution function $f(\mathbf{x},\mathbf{v},t)$; hence we can compute $R(\mathbf{v}'\to\mathbf{v};\mathbf{x},t;\tau)$ under the assumption that one gas molecule of given velocity $\mathbf{v}'$ impinges upon the wall. We shall always assume that we deal with the situation that we have just described; then the kernel $R(\mathbf{v}'\to\mathbf{v};\mathbf{x},t;\tau)$ provides a complete description of the surface scattering law.

If, in addition $\bar{\tau}_e$ is small compared to any characteristic time of interest in the evolution of f, we can let $\tau=0$ in the argument of f appearing in the right hand side of Eq. (2.4), which then becomes:

$$f(\mathbf{x},\mathbf{v},t)|\mathbf{v}\cdot\mathbf{n}| = \int_{\mathbf{v}'\cdot\mathbf{n}<0} R(\mathbf{v}'\to\mathbf{v};\mathbf{x},t)f(\mathbf{x},\mathbf{v}',t)|\mathbf{v}'\cdot\mathbf{n}|d\mathbf{v}' \qquad (\mathbf{x}\in\partial\Omega,\ \mathbf{v}\cdot\mathbf{n}>0) \tag{2.5}$$

where

$$R(\mathbf{v}'\to\mathbf{v};\mathbf{x},t) = \int_0^\infty d\tau\; R(\mathbf{v}'\to\mathbf{v};\mathbf{x},t;\tau) \tag{2.6}$$

Eq. (2.5) is, in particular, valid for steady problems.

Although the idea of a scattering kernel had appeared before, it is only at the end of 1960's that a systematic study of the properties of this kernel appears[7-9]. In particular, the following properties were pointed out[7-13,5]:

1) Non-negativeness, *i. e.* R cannot take negative values:

$$R(\mathbf{v}'\to\mathbf{v};\mathbf{x},t;\tau) \geq 0 \tag{2.7}$$

and, as a consequence:

$$R(\mathbf{v}'\to\mathbf{v};\mathbf{x},t) \geq 0 \tag{2.8}$$

2) Normalization, if permanent adsorption is excluded; *i. e.* R, as a probability density for the totality of events, must integrate to unity:

$$\int_0^\infty d\tau \int_{\mathbf{v}\cdot\mathbf{n}>0} R(\mathbf{v}'\to\mathbf{v};\mathbf{x},t;\tau)d\mathbf{v} = 1 \tag{2.9}$$

and, as a consequence:

$$\int_{\mathbf{v}\cdot\mathbf{n}>0} R(\mathbf{v}'\to\mathbf{v};\mathbf{x},t)d\mathbf{v} = 1 \tag{2.10}$$

3) Reciprocity; this is a subtler property related to the fact that the microscopic dynamics is time reversible and the wall is assumed to be in a local equilibrium state, not significantly disturbed by the impinging molecule. It reads as follows:

$$|\mathbf{v}'\cdot\mathbf{n}|M_w(\mathbf{v}')R(\mathbf{v}'\to\mathbf{v};\mathbf{x},t;\tau) = |\mathbf{v}\cdot\mathbf{n}|M_w(\mathbf{v})R(-\mathbf{v}\to-\mathbf{v}';\mathbf{x},t;\tau) \tag{2.11}$$

and, as a consequence:

$$|\mathbf{v}'\cdot\mathbf{n}|M_w(\mathbf{v}')R(\mathbf{v}'\to\mathbf{v};\mathbf{x},t) = |\mathbf{v}\cdot\mathbf{n}|M_w(\mathbf{v})R(-\mathbf{v}\to-\mathbf{v}';\mathbf{x},t) \tag{2.12}$$

Here M_w is a (non-drifting) Maxwellian distribution having the temperature of the wall, which is uniquely identified apart from a factor.

We remark that the reciprocity normalization relations imply another property:

3') Preservation of equilibrium, *i. e.* the Maxwellian M_w must satisfy the boundary condition (2.4):

$$M_w(\mathbf{v})|\mathbf{v}\cdot\mathbf{n}| = \int_0^\infty d\tau \int_{\mathbf{v}'\cdot\mathbf{n}<0} R(\mathbf{v}'\to\mathbf{v};\mathbf{x},t;\tau)M_w(\mathbf{v}')|\mathbf{v}'\cdot\mathbf{n}|d\mathbf{v}' \qquad (2.13)$$

equivalent to:

$$M_w(\mathbf{v})|\mathbf{v}\cdot\mathbf{n}| = \int_{\mathbf{v}'\cdot\mathbf{n}<0} R(\mathbf{v}'\to\mathbf{v};\mathbf{x},t)M_w(\mathbf{v}')|\mathbf{v}'\cdot\mathbf{n}|d\mathbf{v}' \qquad (2.14)$$

In order to obtain Eq. (2.13) it is sufficient to integrate Eq. (2.11) with respect to $\mathbf{v}'$ and τ, taking into account Eq. (2.9) (with $-\mathbf{v}$ and $-\mathbf{v}'$ in place of $\mathbf{v}'$ and $\mathbf{v}$, respectively. We remark that frequently, one assumes Eq. (2.13) (or (2.14)), without mentioning Eq. (2.11) (or (2.12)); although this is enough for many purposes, reciprocity is very important when constructing mathematical models, because it places a strong restriction on the possible choices. A detailed discussion of the physical conditions under which reciprocity holds has been given by Bärwinkel and Schippers[14].

3. The accommodation coefficients.

The scattering kernel is a fundamental concept in gas-surface interaction, by means of which other quantities should be defined. Frequently its use is avoided by using the so-called accommodation coefficients (a. c.'s), with the consequence of lack of clarity, misinterpretation of experiments, bad definitions of terms and misunderstanding of concepts. The basic information on gas-surface interaction, which should in principle obtained from a detailed calculation based on a physical model as discussed in the next section, is summarized in a scattering kernel. The further reduction to a small set of a. c.'s can be advocated for practical aims, provided this use is firmly related to the scattering kernel.

In order to describe the a. c.'s in a systematic way, one can introduce, for any pair of functions ϕ and ψ, the notations:

$$(\psi,\phi)_+ = \int_{\mathbf{v}\cdot\mathbf{n}>0} \psi(\mathbf{v})\phi(\mathbf{v})M_w(\mathbf{v})|\mathbf{v}\cdot\mathbf{n}|d\mathbf{v} \qquad (3.1)$$

$$(\psi,\phi)_- = \int_{\mathbf{v}\cdot\mathbf{n}<0} \psi(\mathbf{v})\phi(\mathbf{v})M_w(\mathbf{v})|\mathbf{v}\cdot\mathbf{n}|d\mathbf{v} \qquad (3.2)$$

Now, if we factor M_w out of the distribution function f and write:

$$f = M_w \phi \tag{3.3}$$

we can define the a. c. for the quantity ψ when the distribution function at the wall is $M_w\phi$, in the following way:

$$\alpha(\psi,\phi) = [(\psi,\phi)_- - (\psi,\phi)_+]/[(\psi,\phi)_- - (\psi,\iota)_+] \tag{3.4}$$

where ι denotes a constant function, such that

$$(\iota,\iota) = (\iota,\phi) \tag{3.5}$$

Physically the numerator in Eq. (3.4) is the difference between the impinging and emerging flow of the quantity, whose density is ψ, when the distribution is $M_w\phi$; the denominator is the same thing when the restriction of f to $\mathbf{v}\cdot\mathbf{n}>0$ is replaced by the wall Maxwellian, normalized in such a way as to give the same entering flow rate as f. In particular, if we let $\psi = \mathbf{v}\cdot\mathbf{n}$, we obtain the a. c. for normal momentum, if we let $\psi = \mathbf{v}\cdot\mathbf{t}$ we obtain the a. c. for tangential momentum (in the direction of the unit vector $\mathbf{t}$, tangent to the wall), if we let $\psi = |\mathbf{v}|^2$, we obtain the a. c. for energy. It is convenient to restrict the definition in Eq. (3.4) to functions with the property $\psi(\mathbf{v}) = \psi(\mathbf{v}-2\mathbf{n}(\mathbf{n}\cdot\mathbf{v}))$, *i. e.* even functions of $\mathbf{v}\cdot\mathbf{n}$. This condition is not satisfied by $\psi = \mathbf{v}\cdot\mathbf{n}$; accordingly, if one wants to define an a. c. for normal momentum, he has to take $\psi=|\mathbf{v}\cdot\mathbf{n}|$.

In general, $\alpha(\psi,\phi)$ depends upon the distribution function of the impinging molecules; accordingly the definition (3.4) is not so useful, in general. The notion of a. c. becomes more useful, if we select a particular class of functions ϕ, as will be done below.

Let us first investigate the relation between a. c. 's and the scattering kernel $R(\mathbf{v}'\to\mathbf{v})$ (we omit indicating the space and time arguments). We assume that reciprocity, as expressed by Eq. (2.12), holds and define the linear operator A by:

$$A\phi = [M_w(\mathbf{v})|\mathbf{v}\cdot\mathbf{n}|]^{-1}\int_{\mathbf{v}'\cdot\mathbf{n}>0} R(-\mathbf{v}'\to\mathbf{v})M_w(\mathbf{v}')\phi(\mathbf{v}')|\mathbf{v}'\cdot\mathbf{n}|d\mathbf{v}' \tag{3.6}$$

Reciprocity makes A symmetric with respect to the scalar product $(\cdot,\cdot)_+$:

$$(\psi,A\phi)_+ = (A\psi,\phi)_+ \tag{3.7}$$

We now define the reflection or parity operators in velocity space:

$$P\phi = \phi(-\mathbf{v})$$
$$P_n\phi = \phi(\mathbf{v}-2\mathbf{n}(\mathbf{n}\cdot\mathbf{v})) \tag{3.8}$$
$$P_t\phi = \phi(-\mathbf{v}+2\mathbf{n}(\mathbf{n}\cdot\mathbf{v}))$$

and observe that

$$P^2=P_n{}^2=P_t{}^2=I\ ,\ P_tP_n=P_nP_t=P,\quad (\psi,P_t\phi) = (P_t\psi,\phi) \tag{3.9}$$

where I is the identity operator. As stated above, the function ϕ in Eq. (3.4) will be taken to satisfy the restriction

$$P_n\psi = \psi \tag{3.10}$$

Eq. (3.4) then becomes

$$\alpha(\psi,\phi) = [(\psi,P\phi)_+ - (\psi,A\phi)_+]/[(\psi,P\phi)_+ - (\psi,\iota)_+] \tag{3.11}$$

We remark that, once the gas-surface interaction is specified by the kernel $R(\mathbf{v}'\to\mathbf{v})$, then only the impinging distribution function (the restriction of $f=M_w\phi$ to $\mathbf{v}\cdot\mathbf{n}<0$) needs to be known in Eq. (3.11), where, because of Eqs. (3.9) and (3.10), we can replace P by P_t.

Let us now consider particularly meaningful instances of a. c.'s. The first of them, introduced by the Author[7,8], is mathematically simple, but has no direct physical significance for general kernels. In order to define it, let us consider the eigenvalue problem:

$$A\psi = \lambda P_t\psi \tag{3.12}$$

$\psi=\iota$ is an eigensolution corresponding to $\lambda=1$; all the other solutions must correspond to eigenvalues such that $|\lambda|<1$ (unless the kernel $R(\mathbf{v}'\to\mathbf{v})$ is a delta function), thanks to the property[10,12]

$$(A\psi,A\psi)_+ \le (\psi,\psi)_+ \tag{3.13}$$

This inequality follows from the properties of the kernel R and the equality sign holds in two cases only: a)the kernel is a delta function; b)the kernel is general and ψ is constant. Thus, if we discard the singular cases of kernels which reduce to a delta function (which can be easily discussed in a direct fashion), we have together with $\psi_0=\iota$ a sequence of eigenfunctions ψ_k, associated with eigenvalues λ_k (k=1,2,3,...), with $|\lambda_k|<1$ and ψ_k orthogonal to ι (note that $(\iota,P_t\psi_k)=0$ because A is selfadjoint and this implies $(\iota,\psi_k)=0$ because $P_t\iota=\iota$. Then, for any ϕ, we have the a. c. 's

$$\alpha_k=\alpha(\psi_k,\phi) = [(\psi_k,P_t\phi)_+ - (\psi_k,A\phi)_+]/(\psi_k,P_t\phi)_+ = 1-\lambda_k \qquad (k=1,2,3,\dots) \tag{3.14}$$

The obvious advantage of this set of coefficients is that they are independent of the distribution function of the impinging molecules;

further, giving them (together with the set of eigenfunctions ψ_k) is equivalent to giving the scattering kernel. There is also an obvious disadvantage, however: the eigenfunctions ψ_k may be hard to compute and, in most cases, do not possess a simple physical significance.

Another possibility was considered by Shen and Kuščer[15,11,5]. Let us consider a set of physically meaningful quantities $\{\psi_k\}$ and let $\psi = \psi_i$, $\phi = \iota+\phi_k$, where ι is a constant (not necessarily satisfying the constraint in Eq. (3.5); if we still adopt the definition in Eq. (3.4), we obtain a matrix of a. c. 's:

$$\alpha_{ik} = \alpha(\psi_i, \iota+\psi_k) = [(\psi_i, P_t\psi_k)_- - (\psi_i, A\psi_k)_+]/(\psi_i, P_t\psi_k)_- \quad (3.15)$$

These a. c. 's have a clear meaning, but only for the special distribution function $f = M_w(\iota+\psi_k)$, and are easier to compute, in general, than the previous ones, since solving Eq. (3.12) is not needed.

4. Physical models of gas-surface interaction.

The concepts of scattering kernel and a. c. are very useful schemes, where to fit the information arising from a study of the complicated physical phenomena associated with the interaction of a gas molecules with the atoms of a wall. This study is by necessity difficult since we are dealing with a many body problem. Any theoretical model leading to more than purely formal results turns out to be either a crude one or a rough approximation, with the exception of special situations. The support that may come from experimental data is sometimes deceptive, because these data are scanty.

A systematic review of existing models and approximations prior to 1976 may be found in the book by Goodman and Wachman[16]. Here we only sketch a few attempts and indicate a few recent developments.

An early attempt is due to Baule[17], who considered a model of the solid in the form of an array of smooth hard spheres, exercising independent thermal motions; later Logan *et al.*[18] replaced the spheres with hard cubes. Few examples are simple enough to yield to a closed form solution even in the case of a single scattering event. The scattering kernel for this model has been evaluated by Bärwinkel and Schmidt[19]. We remark that the kernels obtained from single scattering events must be cut-off in order to discard the molecules still moving inwards after the interaction. Hence they are not normalized and violate reciprocity, because of the arbitrariness in their way of dealing with multiple scattering events.

Some of the difficulties are avoided by the soft-cube model, proposed by Logan and Keck[20]. In this model each cube carries a rather flat potential well, with the purpose of explaining scattering events at energies of the order of the thermal energy without having to deal with multiple scattering, which is now insignificant. The typical lobular patterns in the angular distributions of scattered molecules, which are seen in experiments, are well reproduced[16]. The model also explains adsorption[20], but does not account for tangential momentum exchange.

A realistic picture can only be obtained by taking an entire piece of solid wall and simulating the dynamics on a computer[21,16], through the so-called trajectory analysis. Recent examples of this kind of calculations are briefly discussed in Sect. 6.

At higher energies the impinging molecule begins to feel the surface waviness, which results into a modified scattering pattern ("structure scattering"). A conspicuous pattern shows two peaks near certain limiting angles (the so called "rainbow effect"[16]).

Models capable of dealing with multiple scattering were proposed by this Author[12,13] but have received little if any attention. They are based on the idea of using a transport equation for the molecule inside the solid, which is regarded as a half-space. Then one has to solve the following equation

$$\mathbf{v}\cdot\frac{\partial P}{\partial \mathbf{x}} + \mathbf{X}(\mathbf{x})\cdot\frac{\partial P}{\partial \mathbf{v}} = LP \qquad (4.1)$$

where P is the probability density of finding the molecule in the half-space which simulates the wall (say $x_1<0$) at position $\mathbf{x}$ with velocity $\mathbf{v}$. $\mathbf{X}$ is the many-body force exerted on the molecule by the solid atoms and L a linear operator (of the Boltzmann and/or Fokker-Planck type) which describes the shorter range interactions. This type of equation is not so easy to deal with, except in artificial but interesting cases; in fact by simulating the many body force by a rigid surface and using collision frequencies (in the Boltzmann-like term) or diffusion coefficients (in the Fokker-Planck term) proportional to v_1, it is possible to recover the Maxwell model and the Cercignani-Lampis model (to be discussed in the next section) for $R(\mathbf{v}'\to\mathbf{v})$. In fact once Eq. (4.1) is solved with the boundary condition:

$$P = \delta(\mathbf{v}-\mathbf{v}') \qquad x_1=0,\ v_1=\mathbf{v}\cdot\mathbf{n}<0 \tag{4.2}$$

then $R(\mathbf{v}'\to\mathbf{v})$ is immediately recovered by:

$$R(\mathbf{v}'\to\mathbf{v})=\frac{|\mathbf{v}\cdot\mathbf{n}|}{|\mathbf{v}'\cdot\mathbf{n}|}\left[P(\mathbf{x},\mathbf{v})\right]_{x_1=0} \tag{4.3}$$

The idea of using a statistical description of the motion of a molecule inside the wall appears in a current approach known as the three-dimensional generalized Langevin model[22,23]. Details of the gas-surface interaction are computed as in a trajectory simulation, but the surface atoms are coupled to the lattice ones through two additional terms in the equation of motion: one of them accounts for the dissipation in the lattice and the second, a fluctuating force, accounts for the effects of lattice vibrations on surface atoms. The lattice atoms, in turn, are harmonically coupled to each other.

Quantum mechanical models have also been developed. Some interesting results concerning elastic scattering have been obtained in the framework of the semiclassical theory by Doll[24,25], Masel *et al.*[26,27], Dion and Doll[28]. The results appear reliable in the case of an incident molecule of high energy and a direction of approach almost normal to the wall. Fully quantum mechanical calculations have also been performed starting with A. Tsuchida[29] and N. Cabrera *et al.*[30,31]. Details on this topic are given in the book by Goodman and Wachman[16] as well as to the paper by Armand and Lapujoulade[32].

5. Mathematical models.

In view of the difficulty of computing the kernel $R(\mathbf{v}'\to\mathbf{v})$ from a physical model of the wall, we shall presently discuss a different procedure, which is less physical in nature. The idea is to construct a mathematical model in the form of a kernel $R(\mathbf{v}'\to\mathbf{v})$ which satisfies the basic physical requirements expressed by Eqs. (2.8), (2.10), (2.12) and is not otherwise restricted except by the condition of not being too complicated.

A possible approach to the construction of such models is through the eigenvalue equation (3.12). Let the λ_k's form a discrete set and let ψ_k be the corresponding eigenfunctions. Then one can show[13] that

$$R(\mathbf{v}'\to\mathbf{v}) = M_w(\mathbf{v})|\mathbf{v}\cdot\mathbf{n}|\sum_{k=0}^{\infty} \lambda_k\phi_k(-\mathbf{v}')\phi_k(\mathbf{v}) \qquad (5.1)$$

where

$$\phi_k(\mathbf{v}) = P_t[\psi_k(\mathbf{v})] \qquad (5.2)$$

In order to construct a model, it is necessary to choose the eigenfunctions ψ_k and the eigenvalues λ_k. Of course, it is convenient to make the choice in such a way that the series in Eq. (5.1) can be evaluated in finite terms. One possibility would be to take only a finite number of eigenvalues different from zero[7,8], but this has the disadvantage that, generally speaking, the positivity requirement, Eq. (2.8), cannot be met. The simplest choice, then, is to take the first eigenvalue to be unity (with a constant eigenfunction, as required by Eq. (2.14)) and the others all equal to the same value $1-\alpha$ ($0\le\alpha\le1$). The kernel then turns out to be

$$R(\mathbf{v}'\to\mathbf{v}) = \alpha M_w(\mathbf{v})|\mathbf{v}\cdot\mathbf{n}| + (1-\alpha)\delta(\mathbf{v}-\mathbf{v}'+2\mathbf{n}(\mathbf{v}'\cdot\mathbf{n})) \qquad (5.3)$$

This is the kernel corresponding to Maxwell model, according to which a fraction $(1-\alpha)$ of molecules undergoes a specular reflection, while the remaining fraction α is diffused with the Maxwellian distribution of the wall M_w. This is the only model for the scattering kernel that appeared in the literature before the late 1960's. Since this model was felt to be somehow inadequate to represent the gas surface-interaction, Nocilla[33] proposed to assume that the molecules are re-emitted according to a drifting Maxwellian with a temperature, which is, in general, different from the temperature of the wall. While this model is useful as a tool to represent experimental data and has been used in actual calculations, expecially in free-molecular flow[34], when interpreted at the light of later developments, it does not appear to be tenable, unless its flexibility is severely reduced[10,35,36]. While the idea of a model like Nocilla's can be traced back to Knudsen[4], the full development of these ideas led to the so-called Cercignani-Lampis (CL) model[10]. From the point of view taken in this section, this model can be easily obtained by taking the eigenfunctions ψ_k to be products of Hermite polynomials in the tangential components of $\mathbf{v}$ times Laguerre polynomials in the square of the normal component $v_n=\mathbf{v}\cdot\mathbf{n}$ of the same vector. The reason of the different treatment of the components will be clear if one thinks that the range of v_n is $[0,\infty)$ rather than $(-\infty,\infty)$ and the weight factor in the natural scalar product $(.,.)_+$ is not $M_w(\mathbf{v})$ but rather $v_n M_w(\mathbf{v})$. The eigenvalues λ_k are taken to be subsequent powers of two parameters. Then the series in Eq. (5.1) can be summed up by means of the so-called Miller-Lebedev formula[10,13] to yield:

$$R(\mathbf{v}'\to\mathbf{v}) =$$

$$= \frac{[\alpha_n\alpha_t(2-\alpha_t)]}{\pi/2}\beta_w^{\,2} v_n \exp\left\{-\beta_w\frac{v_n^2+(1-\alpha_n)v_n'^2}{\alpha_n}-\beta_w\frac{|\mathbf{v}_t-(1-\alpha_t)\mathbf{v}_t'|^2}{\alpha_t(2-\alpha_t)}\right\}\times$$
$$\times I_0\left(\beta_w\frac{2\sqrt{1-\alpha_n}\,v_n v_n'}{\alpha_n}\right) \qquad (5.4)$$

where I_0 denotes the modified Bessel function of first kind and zeroth order defined by

$$I_0(y) = (2\pi)^{-1}\int_0^{2\pi}\exp(y\cos\phi)d\phi \qquad (5.5)$$

The two parameters α_t and α_n have a simple meaning; the first one is the a. c. for the tangential components of momentum, the second one the a. c. for $v_n^{\,2}$, hence for the part of kinetic energy associated with the normal motion. This model became rather popular because it was found by other methods as well: through an analogy with Brownian motion by I. Kuščer *et al.*[36], under a special mathematical assumption by T. G. Cowling[37], through an analogy with the scattering of electromagnetic waves from a surface by M. M. R. Williams[38]. Finally it followed from the solution of the steady Fokker-Planck equation describing a (somewhat artificial) physical model of the wall[12], as was already mentioned in the previous section. Also, the comparison with the data from beam scattering experiments were quite encouraging[10,39,13]. In spite of this, it must be clearly stated that there are no physical reasons why this model should be considered better than others. In particular we remark that any linear combination of scattering kernels with positive coefficients adding to unity is again a kernel that satisfies all the basic properties. Thus, from a kernel with two parameters, such as the CL model we can construct a general model, containing an arbitrary function of those parameters.

We want to mention another method to produce scattering kernels in a simple way. This method was described about 15 years ago[40,13] but does not appear to have been ever used. The starting point is any positive function $K(\mathbf{v},\mathbf{v}')$, defined for $v_n \geq 0$ and $v'_n \geq 0$, symmetric in its arguments and such that such that

$$H(\mathbf{v}') = \int_{\mathbf{v}.\mathbf{n}>0} K(\mathbf{v},\mathbf{v}')M_w(\mathbf{v})v_n d\mathbf{v} \quad (5.6)$$

is not larger than unity for any $\mathbf{v}'$ ($v'_n \geq 0$). Then we form the kernel

$$R(\mathbf{v}'\to\mathbf{v}) = v_n M_w(\mathbf{v})K(\mathbf{v},-\mathbf{v}') + v_n M_w(\mathbf{v})\frac{(1-H(-\mathbf{v}'))(1-H(\mathbf{v}))}{\int(1-H(\mathbf{w}))w_n M_w(\mathbf{w})d\mathbf{w}} \quad (5.7)$$

In fact in this way both reciprocity and normalization are clearly satisfied and positivity follows from the assumpion on $H(\mathbf{v})$.

In order to illustrate this procedure we present the costruction of a model chosen to be as close as possible to Nocilla's model and still satisfy the requirements expressed by Eqs. (2.8), (2.10) and (2.12). We start from the assumption (to be corrected later) that the emerging distribution is a drifting Maxwellian, when the imping distribution is a a delta function. Then the candidate kernel is:

$$R_0(\mathbf{v}'\to\mathbf{v}) = A'|v'_n|^{-1}v_n \exp(-\beta'|\mathbf{v}-\mathbf{u}'|^2) \quad (5.8)$$

where $A'=A(-\mathbf{v}')$, $\beta=\beta(-\mathbf{v}')$ and $\mathbf{u}'=\mathbf{u}(-\mathbf{v}')$ can be chosen to be dependent on $\mathbf{v}'$. We satisfy reciprocity first; this requires

$v_n\exp(-\beta_w|\mathbf{v}'|^2)A'\exp(-\beta'|\mathbf{v}-\mathbf{u}'|^2)=|v_n'|\exp(-\beta_w|\mathbf{v}|^2)A\exp(-\beta|\mathbf{v}'+\mathbf{u}|^2)$

where $A=A(\mathbf{v})$, $\beta=\beta(\mathbf{v})$ and $\mathbf{u}=\mathbf{u}(\mathbf{v})$. From this equation we conclude that the only resonable choice is $\mathbf{u}(\mathbf{v}) = a\mathbf{v}_R$ ($\mathbf{v}_R=\mathbf{v}-2\mathbf{n}(\mathbf{v}\cdot\mathbf{v})$) and $\beta(\mathbf{v})= b$ (a and b constants). Then

$$\beta_w|\mathbf{v}'|^2)+b|\mathbf{v}+a\mathbf{v}_R'|^2=\beta_w|\mathbf{v}|^2)+b|\mathbf{v}'+a\mathbf{v}_R|^2 \quad (5.9)$$

and $\beta_w+ba^2=b$ or $b=\beta_w(1-a^2)^{-1}$. Thus $A'=|v_n'|$

$$R_0(\mathbf{v}'\to\mathbf{v}) = cv_n\exp(-\beta_w(1-a^2)^{-1}|\mathbf{v}-a\mathbf{v}_R'|^2) \quad (5.10)$$

where c is a constant.

The symmetric kernel $K(\mathbf{v},\mathbf{v}')$ associated with this choice is

$$K(\mathbf{v},\mathbf{v}')= c\exp(-\beta_w(1-a^2)^{-1}|\mathbf{v}+a\mathbf{v}_R'|^2+\beta_w|\mathbf{v}|^2) \quad (5.11)$$

If we now calculate $H(\mathbf{v})$ as defined in Eq. (5.3), we obtain

$$H(\mathbf{v}') = \int_{\mathbf{v}.\mathbf{n}>0} K(\mathbf{v},\mathbf{v}')M_w(\mathbf{v})v_n d\mathbf{v} = (c\pi/2)\beta_w^{-2}(1-a^2)^2[\exp(-a^2 v_n'^2 \beta_w (1-a^2)^{-1} + a v_n' \beta_w^{-1/2}(1-a^2)^{1/2}\sqrt{\pi}\ \mathrm{erfc}(-a v_n' \beta_w^{1/2}(1-a^2)^{-1/2})] \quad (5.12)$$

where erfc is the complementary error function defined by

$$\mathrm{erfc}(y) = \frac{2}{\sqrt{\pi}}\int_y^\infty \exp(-t^2)dt \quad (5.13)$$

Then the model is completely evaluated. The kernel reads as follows:

$$R(\mathbf{v}'\to\mathbf{v}) = c v_n \exp(-\beta_w (1-a^2)^{-1}|\mathbf{v}-a\mathbf{v}_R'|^2) + v_n \exp(-\beta_w|\mathbf{v}|^2)\frac{(1-H(-\mathbf{v}'))(1-H(\mathbf{v}))}{\int(1-H(\mathbf{w}))w_n\exp(-\beta_w|\mathbf{w}|^2)d\mathbf{w}} \quad (5.14)$$

where $H(\mathbf{v})$ is given by Eq. (5.12). The integral in the denominator can be easily evaluated to give

$$\int(1-H(\mathbf{w}))w_n M_w(\mathbf{w})d\mathbf{w} = (\pi/2)\left[\beta_w^{-2} - c(\pi/2)\beta_w^{-4}\left((1-a^2)^4 + a(1-a^2)^{3/2}(\pi-\cos^{-1}a) + a^2(1-a^2)^2]\right)\right] \quad (5.15)$$

As far as we know, this model has not been published before.

All these models contain pure diffusion according to a nondrifting Maxwellian as a limiting case. The use of this model is justified for low-velocity flows over technical surfaces, but is inaccurate for flows with orbital velocity. In fact, the elaboration of the measurements of lift-to-drag ratio for the Shuttle Orbiter in the free-molecular regime[36] implies a significant departure from diffuse, fully-equilibrated re-emission of molecules at the wall.

6. Testing models.

The traditional methods of testing a model for gas surface interactions is to compare with molecular beam experiments and drag and heat transfer measurements in free molecular flow. The first type of test is, in principle, the best one can think of, but requires a lot of systematic work, which does not appear to have been done. Also the information from molecular beam experiments can be overwhelming for certain applications, such as drag, lift and heat transfer calculations, and one must think carefully about how the amount of material arising from experiments should be represented and used. Good recent surveys of experimental work and its comparison with existing theories are due to Hurlbut[37,38].

An early example of a partial success of trajectory calculations is also due to Hurlbut[21]. Recently, there have been several calculations of this kind that shed light on gas surface interaction. In particular Hulpke and Mann[39] and Tenner *et al.*[40-42] conducted studies on the scattering of potassium ions from clean tungsten surfaces and analysis of ion trajectories and found reasonable agreement with experiment. Among the important result arising from these studies, one can quote the following ones: a) trajectory analysis can yield simulations representing details of the physical scattering process very accurately; b) three-dimensional analyses are required, because two dimensional analyses miss the lower energy tail in the energy distribution of scattered particles; c) the exact value of the parameters in the interaction potential can be fixed only after a comparison with experiments is made.

Another example of trajectory modelling accompanying experimental data is provided by the paper of Kolodney *et al.*[43], who considered mercury scattering from magnesium oxide.

With a few exceptions, scattering studies have used single crystal surfaces as targets. Hurlbut[37] made the important remark that, while the use of these well-defined surfaces is essential for progress in surface science, one should, in order to understand the interaction of the upper atmosphere with the surfaces of orbiting shuttles, study scattering events and the attendant physics and chemistry on surfaces with the appropriate composition.

An experimental validation of the CL model was attemped by Bellomo, Dankert, Legge and Monaco[44]. According to these authors, the CL model does not reproduce the experimental drag and the heat flow on a metal plate placed in a supersonic nearly free-molecular flow. The interpretation of the experimental results is, however, based on a misinterpretation of the CL model; in fact the above-mentioned authors try to fit the experimental data for the drag and heat transfer coefficients *versus* the plate temperature by means of expressions obtained from the CL model under the additional tacit assumption that the parameters in the model are temperature independent. The only firm conclusion that one can draw from the experimental data is that the recovery temperature does not depend on the angle of attack θ (at least in the range of the experiments, $45^0 \leq \theta \leq 90^0$); this is known[45] to imply that the accommodation coefficients of the tangential and normal kinetic energies must be equal, under the envisaged experimental conditions.

With the recent progress achieved in the numerical solution of the Boltzmann equation by means of more and more powerful computers, it is now possible to think about more sophisticated comparisons in the transition regime. As pointed out by Becker *et al.*[46], their measurements provide such an opportunity. This prompted several authors[47-49] to compute the flow experimentally investigated by Becker *et al.*[46]. Thus, *e. g.* , the computations in Ref. 48 lead to the conclusion that neither the Maxwell model nor the CL model can reproduce the whole set of experimental results with sufficient accuracy. In particular, temperature and density are well reproduced but the velocity field is not. Similar conclusions were also drawn by Hurlbut[47]. Apparently one needs a gas-surface interaction model that assigns perfect accommodation to temperatures but gives rather high slip velocities. New models such as the one given by Eq. (5.9) or others to be built according to the method discussed in Sect 5 should be used for further comparisons.

REFERENCES

1. A. Kundt and E. Warburg, *Pogg. Ann. Phys.* **155**, 337 (1875).
2. J. C. Maxwell, *Phil. Trans. Royal Soc.* **I**, Appendix (1879).
3. M. S. v. Smoluchowski, *Phil. Mag.* **46**, 192 (1898).
4. M. Knudsen, *The Kinetic Theory of Gases*, Methuen, London (1950).
5. I. Kuščer, in *Fundamental Problems in Statistical Mechanics* **IV**, E. G. D. Cohen and W. Fiszdon, Eds., Osssolineum, Warsaw (1978).
6. J. H. de Boer, *The Dynamical Character of Adsorption*, Clarendon Press, Oxford (1953).
7. C. Cercignani, in *Transport Theory*, G. Birkhoff *et al.*, eds., SIAM-AMS Proceedings, Vol. I, p.249, Providence (1968).
8. C. Cercignani, *Mathematical Methods in Kinetic Theory*, Plenum Press, New York (1969).

9. I. Kuščer, in *Transport Theory Conference*, AEC Report ORO-3588-1, Blacksburgh, Va. (1969).
10. C. Cercignani and M. Lampis, *Transport Theory and Stat. Phys.* **1**, 101 (1971).
11. I. Kuščer, *Surface Science* **25**, 225 (1971).
12. C. Cercignani, *Transport Theory and Stat. Phys.* **2**, 27 (1972).
13. C. Cercignani, *The Boltzmann Equation and its Applications*, Springer, New York (1987).
14 K. Bärwinkel and S. Schippers, in *Rarefied Gas Dynamics: Space-Related Studies* E. P. Muntz, D. P. Weaver and D. H. Campbell, eds., p. 487, AIAA, Washington (1989).
15. S. F. Shen and I. Kuščer, in *Rarefied Gas Dynamics*, Dini *et al.* eds.,vol. I, 109 (1974).
16. F. O. Goodman and H. Y. Wachman, *Dynamics of Gas-Surface Scattering*, Academic Press, New York (1976).
17. B. Baule, *Ann. d. Phys.* (4) **44**, 145 (1914).
18. R. M. Logan, R. E. Stickney, *J. Chem. Phys.* **44**, 195 (1966).
19. K. Bärwinkel and H. J. Schmidt, in *Rarefied Gas Dynamics*, S. S. Fischer, ed., Vol. I, 150, AIAA, New York (1981).
20. R. M. Logan and J. C. Keck, *J. Chem. Phys.* **49**, 860 (1968).
21. F. C. Hurlbut, in *Rarefied Gas Dynamics*, M. Becker and M. Fiebig, eds. Vol.II, Ax-E. 3-1, DFVLR, Porz-Wahn (1974).
22. J. G. Tully, *J. Chem. Phys.* **73**, 1975 (1980).
23. J. G. Tully, *Annual Review of Physics and Chemistry* **31**, 319 (1980).
24. J. D. Doll, *Chem. Phys.* **3**, 257 (1974).
25. J. D. Doll, *J. Chem Phys.* **61**, 954 (1974).
26. R. I. Masel, R. P. Merril and W. H. Miller, *Surf. Sci.* **46**, 681 (1974).
27. R. I. Masel, R. P. Merril and W. H. Miller, *J. Chem. Phys.* **64**, 45 (1976).
28. D. R. Dion and J. D.Doll, *Surf. Sci.* **58**, 415 (1976).
29. A. Tutschida, *Surf. Sci.* **14**, 375 (1969).
30. N.Cabrera, V.Celli, J.R. Manson, *Phys.Rev.Lett.* **22**, 346 (1969).
31. N. Cabrera, V. Celli, F. O. Goodman and J. R. Manson,*Surf. Sci.* **19**, 67 (1970).
32. G. Armand and J. Lapujoulade, in *Rarefied Gas Dynamics*, R. Campargue ed., vol. II, 1329, CEA, Paris (1979).
33. S. Nocilla, in *Rarefied Gas Dynamics*, L. Talbot, ed., 169, Academic Press, New York (1961)
34. F. Hurlbut and F. S. Sherman, *Phys. Fluids* **11**, 486 (1968).
35. M. N. Kogan, *Rarefied Gas Dynamics*, Plenum Press, New York (1969).
36. I. Kuščer, J. Možina and F. Krizanic, in *Rarefied Gas Dynamics*, D. Dini *et al.*, eds., Vol. I, 97, Editrice Tecnico-Scientifica, Pisa (1974).
37. T. G. Cowling, *J. Phys. D: Appl. Phys.* **7**, 781 (1974).
38. M. M. R. Williams, *J. Phys. D: Appl. Phys.* **4**, 1315 (1971).
39. C. Cercignani, in *Rarefied Gas Dynamics*, D. Dini *et al.* eds.,vol. I, 75, Editrice Tecnico- Scientifica, Pisa (1974).
40. C. Cercignani, *Theory and Application of the Boltzmann Equation*, Scottish Academic Press, Edinburgh and Elsevier, New York (1975).
41. R. C. Blanchard, Paper No. ICAS 86-2.10.1, 15th ICAS Congress, London (September 1986).

42. F. C. Hurlbut, in *Rarefied Gas Dynamics: Space-Related Studies*, E. P. Muntz, D. P. Weaver and D. H. Campbell, eds., p. 419, AIAA, Washington (1989).
43. F. C. Hurlbut, *Thermophysical Aspects of Re-entry Flows*, J. N. Moss and C. D., Scott eds., AIAA, New York (1986).
44. E. Hulpke and K. Mann, *Surf. Sci.* **133**, 171 (1983).
45. A. D. Tenner, K. T. Gillen, T. C. M. Horn, J. Los and A. W. Klein, *Phys. Rev. Letters* **52**, 2183 (1984).
46. A. D. Tenner, R. P. Saxon, K. T. Gillen, D. E. Harrison, T. C. M. Horn and A. W. Klein, *Surf. Sci.* **172**, 121 (1986).
47. A. D. Tenner, K. T. Gillen, T. C. M. Horn, J. Los and A. W. Klein, *Surf. Sci.* **172**, 90 (1986).
48. E. Kolodney, A. Amirav. R. Elber,and R. B. Gerber, *Chem. Phys. Letters* **113**, 303 (1985).
49. N. Bellomo, C. Dankert, H. Legge, in *Rarefied Gas Dynamics*, O. M. Belotserkovskii, M. N. Kogan, S.S. Kutateladze, A. K. Rebrov, eds., Vol. I, p. 421, Plenum Press, New York (1985).
50. C. Cercignani and M. Lampis, *Jour. Appl. Math. and Phys. (ZAMP)* **27**, 733 (1976).
51 M. Becker, F. Robben and R. Cattolica, *AIAA Journal* **12,** 1247 (1974).
52. F. C. Hurlbut, *AIAA Paper 87-1545* (1987).
53. C. Cercignani and A. Frezzotti, in *Rarefied Gas Dynamics: Theoretical and Computational Techniques*, E. P. Muntz, D. P. Weaver and D. H. Campbell, eds., p. 552, AIAA, Washington (1989).
54. W.L. Hermina, *AIAA Paper 87-1547* (1987).

Computational Fluid Dynamics for Hypersonic Airbreathing Aircraft

James L. Thomas
Douglas L. Dwoyer
Ajay Kumar
NASA Langley Research Center, Hampton, Virginia, USA

Abstract

Aspects of the requirements placed on the development and use of computational fluid dynamics for the analysis and design of hypersonic airbreathing aircraft are discussed. Current capabilities in terms of physical models and algorithms are assessed and an example of the code validation process for a strong-interaction hypersonic flow field is discussed in some detail. Several examples of advanced applications are also presented.

Introduction

Research in the general area of hypersonics was almost nonexistent in the period from the early 1970's to the mid 1980's.[1] During this period, analytical research directed specifically at hypersonics virtually disappeared. There were only a few pockets of research activities located primarily in the United States at the National Aeronautics and Space Administration (NASA) centers. However, in the past few years there has been an explosion of activities in this area including government research laboratories, universities, and industries. This substantially increased interest in hypersonics has been caused in part by the need for developing a new class of vehicle, called the aero-space plane, powered by an airbreathing propulsion system.

A hypersonic airplane powered by airbreathing engines can gain a performance advantage over a rocket powered vehicle by using the atmosphere as the oxidizer. In order to achieve or realize this advantage, however, these airbreathing engines require large capture areas and must be closely integrated with the airframe to avoid excessive drag penalties. This requirement has resulted in designs in which almost the entire undersurface becomes part of the propulsion system (Fig. 1). The forebody precompresses the flow before it enters the engine inlets and the flared afterbody acts as part of the nozzle by further expanding the exhaust products from the engine. The engine itself may have features such as cut-back cowls, spill windows or doors, and swept compression surfaces to operate it over a wide Mach number range. Struts or centerbodies may be necessary to provide additional compression in the inlets and locations for fuel injections in the combustors. A large fraction of the net thrust at hypersonic speeds is derived from forces acting on the forebody and

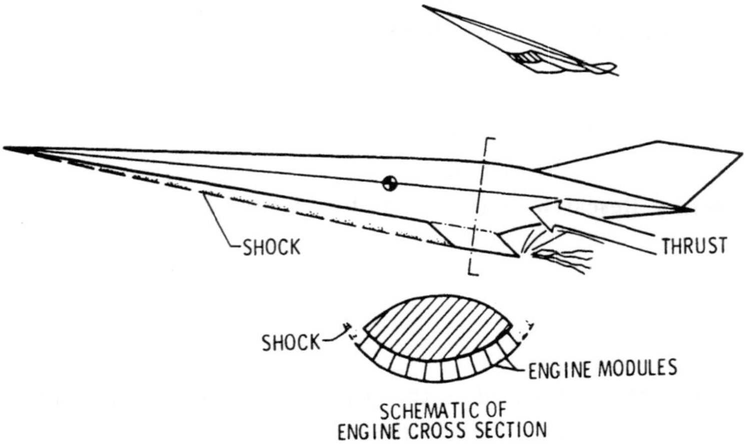

FIGURE 1. AIRFRAME-INTEGRATED SCRAMJET-ENGINE CONCEPT.

afterbody of the vehicle.[2] Thus the analysis and design must include from the outset the combustor and airframe simultaneously rather than separately as in traditional aircraft engine airframe integration problems.

Due to lack of available ground facilities for testing at high Mach numbers (the ones that are available do not simulate the flight enthalpies at these Mach numbers) and due to the complex and integrated nature of the flow field, traditional wind-tunnel based design techniques are not adequate. Modern computational fluid dynamics (CFD) technology offers a promising alternative means of analyzing, designing, and optimizing the hypersonic vehicles. It will be used to complement and extend wind tunnel model results to full-scale flight conditions where only a limited number of tests can be done and will be the dominant analysis technique at the high Mach numbers. However, to do so requires further development in CFD technology, new or improved physical models, and a major focus on large-scale computations. Critical high Mach number problem areas are propulsion-airframe integration, propulsion component design and optimization, and the prediction and understanding of transition, turbulence, mixing, and supersonic combustion.

Through the 1970's and early 1980's research in hypersonic internal flows focused on the development of the individual module of a supersonic combustion ramjet (scramjet) engine for airbreathing propulsion. Since this research took place when CFD was rapidly maturing, computational techniques were developed for analyzing scramjet components. Ref. 3 provides a comprehensive survey of the CFD techniques for analyzing hypersonic internal flows and indicates that computational codes are fairly mature for inlets but somewhat less so for combustors and nozzles.

Interest in hypersonic external aerodynamic predictions throughout the 1970's focused on entry and reentry vehicles. A number of sophisticated codes based on the viscous shock layer (VSL) approximation as well as parabolized Navier-Stokes (PNS) and Reynolds-averaged Navier-Stokes (RANS) equations were developed that included real gas, radiation, ablation, and wall catalysis effects.[4] Thus, it is apparent that CFD codes based on PNS and RANS equations exist today that can analyze flow over hypersonic configurations and through the engine components. However, it is necessary to extend the current CFD technology to one of the most crucial problems in the development of the hypersonic airbreathing airplane – propulsion/airframe integration. To integrate the airframe and propulsion system at high Mach numbers, CFD must treat flow fields substantially more complex than those around the entry vehicles. Ref. 5 contrasts the differing CFD requirements for hypersonic airplanes versus entry vehicles.

The present paper reviews the status of CFD for hypersonic flows in terms of algorithms, their capability to model the relevant physics, and their application to analysis/design optimization of hypersonic configurations. Since CFD simulations must be used heavily in the design of a hypersonic airbreathing vehicle, it is critical that the computer codes in use be validated for their intended application, An example of the code validation process is given for the highly separated hypersonic flow over a compression ramp. Some examples of advanced CFD applications are then presented.

Physical Models

The successful validation of a CFD code to a particular application hinges on the degree to which the governing equations and physical models (i.e. chemistry and turbulence models) represent the flow being considered and the influence of numerical error in the computed solutions. Because of the requirement to optimize the vehicle performance over a range of Mach numbers, a variety of physical models are necessary. The flow around a hypersonic airplane is predominantly three-dimensional and is dominated by viscous effects. Real gas effects become important for flight Mach numbers beyond about ten. Some of the critical design issues are shown in Fig. 2 for a hypersonic airplane concept. Due to the highly integrated nature of vehicle design, there is a strong coupling between the vehicle and propulsion system flow fields. It is necessary to predict accurately not only the state of the boundary layer developed on the forebody but also the entire shock-layer profile in order to be able to determine the mass and momentum flux entering the inlets. Trade studies for candidate U.S. Aero-Space Plane concepts have shown a nearly one-to-one relationship between percentage decreases in inlet mass capture to percentage increases in the take-off gross weight.

After the inlet face, the geometric complexity of the flow-field boundaries increases dramatically. At off-design conditions, the inlets will spill a substantial amount of flow thereby setting up a complex multiple inlet forebody flow-field interaction. The aft end of the hypersonic airplane is dominated by the interaction of the multiple internal/external nozzle system flow with the vehicle wing/body flow. The nozzle flow includes real gas effects throughout the operating envelope of the vehicle and, at off-design conditions, possible flow separation on the external nozzle surface may occur. Requirements for accu-

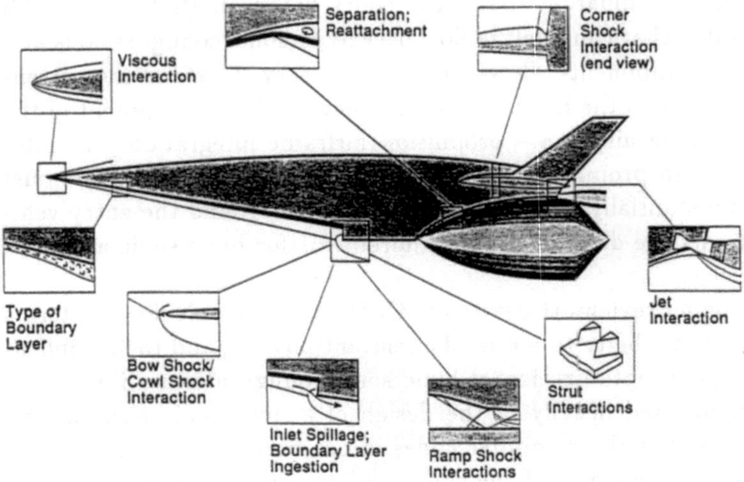

FIGURE 2. DESIGN ISSUES FOR A HYPERSONIC AIRBREATHING AIRCRAFT.

rate prediction of the complete three-dimensional flow field are again important on the external nozzle in order to estimate the nozzle thrust coefficients and the direction of the net thrust vector. The real-gas models incorporated in the nozzle flow analysis must include the chemistry of the combustor products as well as air chemistry.

The combustor flow field presents perhaps the most difficult situation for the CFD codes both in terms of physical modeling as well as flow complexity. Important physical/chemical processes in the combustor include fuel/air mixing, ignition, combustion, and shock turbulence/kinetics interactions. One of the major concerns in the high-speed combustors is to enhance the mixing of fuel and air which is adversely affected with increasing Mach numbers.[6] There are a number of locations in a typical combustor where highly detailed analysis of very localized processes is required, such as in the neighborhood of fuel injectors and flameholders. Refs. 7-10 describe some of the codes developed for combustor flow-field analysis.

A major issue in CFD code development for hypersonic airplane is transition and turbulence modeling. It is extremely important to know the transition location on the vehicle and its extent since transition ambiguity can significantly increase the gross weight of the vehicle. The use of a turbulent rather than laminar boundary layer model, for instance, has a significant effect on the extent of flow separation at compression surfaces, the mass flow characteristics at the inlet, and fuel mixing and combustor efficiency. These changes lead to radically different material selections, thermal management systems, and mission profiles. So far as turbulence modeling for high Mach numbers is concerned, it appears that the existing models are adequate for the attached boundary layer but significant developmental work is required for predicting turbulent mixing in free-shear layers, shock-turbulent boundary-layer interaction, and three-dimensional separation. Real-gas effects and compressibility effects at high Mach numbers need to be incorporated in the models. The kinetics/turbulence interaction is of crucial importance for combustor flow and current models are not well-established. At high altitudes, the flow over the vehicle goes from turbulent to laminar and low-density effects may become important. Transition and turbulence models must, therefore, account for this relaminarization process.

It is apparent from the preceding discussion that there are a number of shortcomings in the physical/chemical modeling area and significant effort is necessary to model the physics of high Mach number flow. The physics that need to be modeled changes significantly over the flight envelope; CFD algorithms should be flexible enough to account for the changing physics if these codes have to be used for the entire range of flight conditions.

Computational Algorithms

As mentioned earlier, the flow field around a hypersonic airplane is three-dimensional and is dominated by viscous effects. It is, therefore, necessary to include viscous terms in the governing equations. In general, the codes that are being developed or used today are based on either the PNS or RANS equations. Fig. 3 shows the various regions where particular algorithms could be used for the simulation of a hypersonic airplane flow field. Insofar as applications are concerned, the PNS codes are the least expensive of the currently available methods for computing hypersonic airplane flow fields. Due to their inability to predict streamwise separation, however, their application is mostly restricted to external flows with limited use to internal flows. Near design conditions, the PNS algorithms could be used on the forebody and in the nozzle and afterbody regions.

The governing equations are generally solved by either central-difference algorithms, upwind-difference algorithms, or spectral methods. Of these, the codes that use the central-difference algorithms are the most mature and have been extended to include the most complete physical/chemical models (e.g. Refs. 7-18). The introduction and use of upwind-difference based codes (e.g. Refs. 19-21) has been spurred by the need for improved shock capturing capabil-

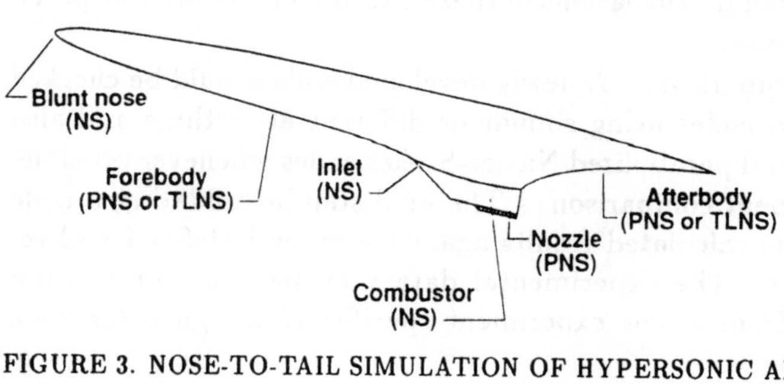

FIGURE 3. NOSE-TO-TAIL SIMULATION OF HYPERSONIC AIRPLANE.

ity in hypersonic flows. An extensive effort has been underway to include more advanced physical and chemical models in these codes. These algorithms are now in wide use in production codes and have successfully predicted complex internal and external flow fields for hypersonic configurations. The spectral methods have proved extremely useful in providing high resolution solutions of linear stability calculations and for direct simulations of turbulence at low Reynolds numbers for simple geometries, such as channels and mixing layers. Recent work has been directed toward using the direct simulations to provide guidance for the development of turbulence models at high Mach numbers.

The CFD codes can be used either for analysis or design/optimization of a configuration. At the present time, use of the RANS codes is primarily restricted to the analysis of the flow. Because of the large computer resource requirements, even a 2-D Navier-Stokes algorithm is considered too expensive and time consuming to be used in design. However, codes based on the PNS method are certainly finding their use in the design process. Ref. 22 provides a good study of CFD code applications to practical problems. A tremendous amount of information can be generated by a CFD code and use of sophisticated graphics/post-processing tools is required to extract meaningful information about the flow from the large data base.

Validation

Before using a CFD code for an applied program, a necessary step is to validate it. Code validation requires detailed and/or global experimental data of high quality taken in well calibrated wind tunnels. The model geometry should be well defined, error bounds on the data should be specified, and the type of flow (laminar, transitional, or turbulent) should be known.

Code validation over the entire operating flight envelope requires validating the code over small ranges of flight envelope. As the vehicle goes through the flight envelope, the physics that needs to be resolved changes. There are three steps involved in the validation process:

(a) Consistency and internal checks. - The code should be checked for mass, momentum, and energy conservation and compared against available exact solutions for inviscid and viscous problems. Frequently, there are internal numerical control parameters in a code which can affect calculated results. A study of the effects of small perturbations of these parameters on the computed results should also be made.

(b) Code to code comparison. - A newly developed code should be checked against other established codes using similar or different algorithms and also against boundary-layer and parabolized Navier-Stokes codes whenever possible.

(c) Code to experiment comparison. - The final step in validating a code consists of comparing the calculated results against some well-defined and reliable experimental data. The experimental data may be from an existing experiment or may be from a new experiment specifically designed for code validation.

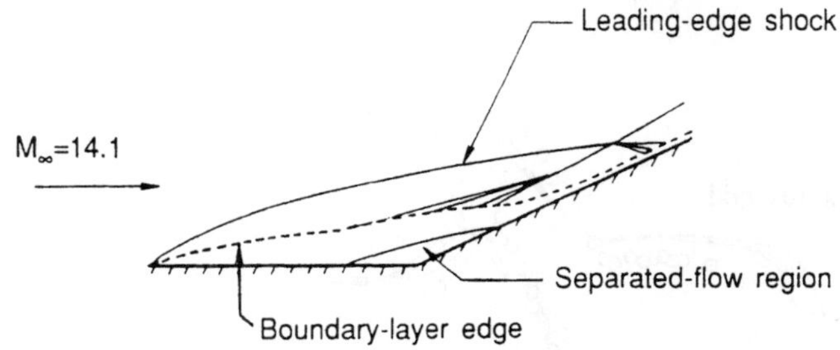

FIGURE 4. HYPERSONIC COMPRESSION CORNER FLOWFIELD.

It is important to realize that due to a lack of activities in the area of hypersonics for over a decade, a vacuum exists in the available experimental data that can be used for code validation. The experimental facilities and required high-quality instrumentation need to be developed so that a reliable experimental data base can be generated. In fact, there seems to be a general consensus that in the coming few years, most of the high-speed testing will be done to produce the extensive data base necessary for code validation. Particular attention must be paid to obtaining flow field surveys and experimental boundary condition measurements in order for the results to be useful for code validation.[23]

One particular validation study is discussed below – the hypersonic flow over a compression corner formed by the intersection of a flat plate and a 24-degree wedge. Shown schematically in Fig. 4, the flow separates upstream of the corner region as a result of the strong interaction of the boundary layer with the hypersonic external flow. Reattachment occurs downstream of the ramp where the boundary layer thins rapidly due to the compression, resulting in large increases in skin friction and heat transfer. The flowfield has significant viscous/inviscid interactions typical of those found in the flow field within the propulsion system of a hypersonic vehicle.

Extensive experimental data measured by Holden and Moselle[24] in the Calspan 48-inch Shock Tunnel are available. The nominal flow conditions are $M_\infty = 14.1$, $T_\infty = 160^oR$, and $Re_L = 1\text{x}10^5$, where L is the length of the flat-plate portion of the model. The Reynolds number is low enough that the flow remains laminar. Furthermore, even though the free-stream Mach number was high, the free-stream temperature was low enough that there were no significant real-gas effects. In the experiment, values of surface pressure, skin friction, and heat transfer were measured in the centerplane of the model which had a spanwise length that was thought to be sufficient to produce two-dimensional flow in the measurement region.

This test case has been and continues to be widely used in computational studies.[22,25,26] The first computation using the full Navier-Stokes equations for this case was made by Hung and MacCormack.[25] They obtained good agreement with the experimental data for 15- and 18-degree wedge angles, but their solution significantly underpredicted the size of the separated-flow region for the 24-degree wedge.

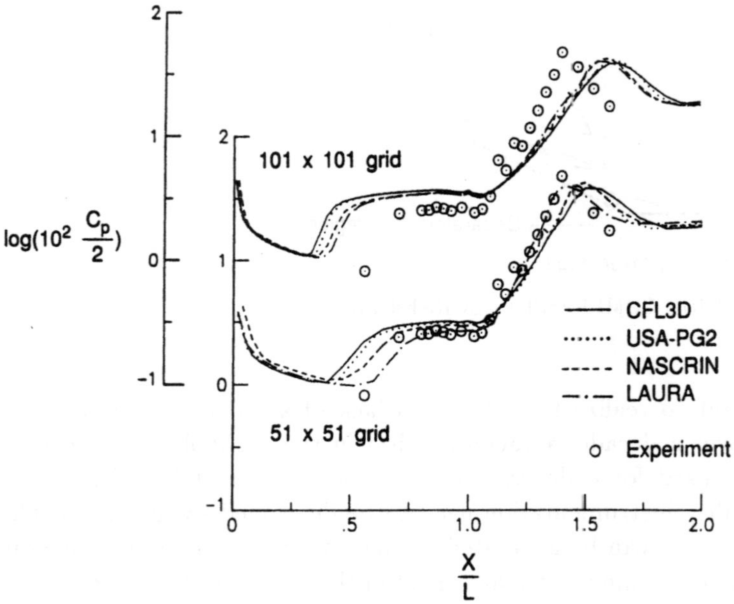

FIGURE 5. COMPARISON OF COMPUTATION AND EXPERIMENT FOR 24-DEGREE WEDGE COMPRESSION.

Comparisons were made between the computations from four different codes and experimental data for the 24-degree wedge. Three of the codes (CFL3D,[19] USA-PG2[20], and LAURA[21]) use recently-developed upwind differencing technology for the convective and pressure terms. The fourth code (NASCRIN) uses the explicit central-difference method of MacCormack.[14–15] Computations on a series of grids (51x51 and 101x101) were used to assess the effect of truncation error on the accuracy.

Fig. 5 shows a comparison of the computed surface-pressure coefficient with experimental values for all four codes. The four codes predict different extents of separation even with the 101 x 101 grid. The effect of refining the grid is to increase the size of the extent of separation, as judged by the extent of the plateau pressure. As a result of the larger separation extent, the shock interaction is altered, moving the peak value of pressure on the ramp downstream in comparison to the experiment. Note that since the size of the separation increases with grid refinement, it is possible to match the experimental data for any one of the four codes with an arbitrarily selected grid density; hence it is possible without extensive grid refinement to be misled in the validation process.

It was suspicioned that perhaps the experimental data was obtained before steady flow had been fully established during the short run time available in the shock tunnel. The flow in the experiment reached a steady state in approximately 4 milliseconds (ms), and the total run time was approximately 10 ms. In order to resolve this question, time-accurate calculations were made with three of the codes. Fig. 6 shows typical computations obtained at five intermediate

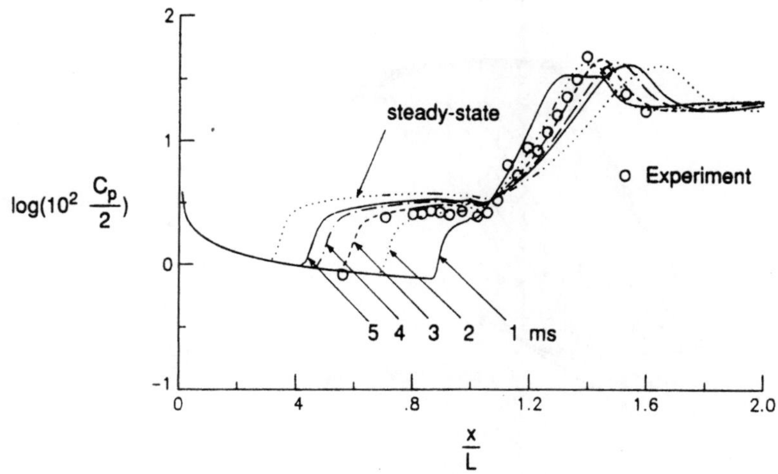

FIGURE 6. TIME EVOLUTION OF THE 2-D PRESSURE VARIATION.

times between 1 and 5 ms. The separated-flow region is predicted reasonably well at a point in time between 2 and 3 ms, but the size of the region continues to increase as the solution is further advanced in time. It took more than 12 ms to establish steady flow in the computations. Additional computations for other wedge angles indicated that it required significantly more time than 4 ms for the computed flow field to reach its steady-state even for a separated-flow region with a size comparable to that found in the experiments.

The experimental data for which comparisons are shown were made on the center section of a plate of spanwise width of two feet with no side plates to constrain the flow. The large pressures induced on the flat plate and ramp can cause a spanwise flow of the low-momentum fluid in the separated zone, which tends to provide a three-dimensional relief effect and thereby decrease the extent of separation. Results by Putnam[27] and Ball[28] are available which demonstrate experimentally this effect. In order to simulate this effect computationally, three-dimensional calculations were made with a grid which extended from the centerline of the model past the side edge. Symmetry conditions were used at the center section and approximate supersonic outflow boundary conditions were used at the sides of the computational domain outside of the ramp surface. Calculations were made with two different grids, each of which had 25 points in the spanwise direction, 19 of which were on the plate.

Fig. 7 shows results from the calculation with a 101 x 101 grid in each streamwise plane. The streamlines in the flow very near the model surface are visualized using particle traces. The separation and reattachment lines show that the size of the separated-flow region decreases across the plate from the centerplane to the edge. The pressure contours in the downstream plane on the ramp indicate an expansion of the flow in the spanwise direction near the edge of the plate to reduce the pressure to the free-stream value. A comparison of the computed 2-D and 3-D centerplane surface-pressure distributions with the experimental data is shown in Fig. 8, including a 2-D solution using a 201 x

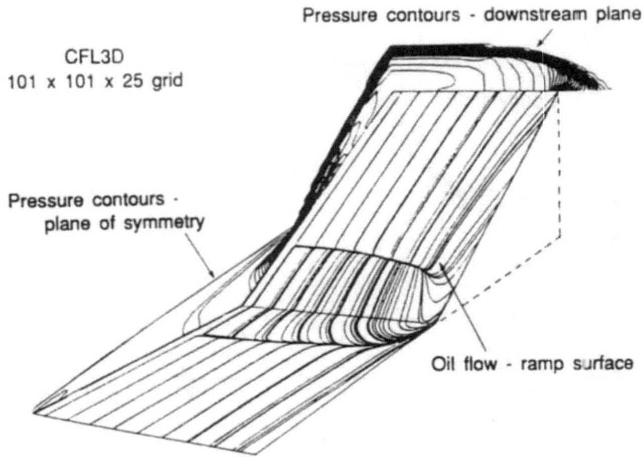

FIGURE 7. SURFACE STREAMLINES AND PRESSURE CONTOURS FROM 3-D COMPUTATION OF HYPERSONIC COMPRESSION CORNER.

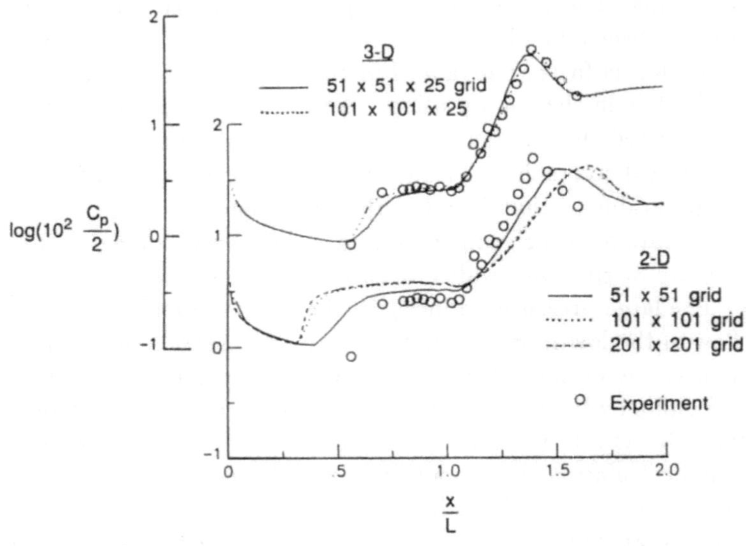

FIGURE 8. THREE-DIMENSIONAL EFFECTS IN THE HYPERSONIC COMPRESSION CORNER FLOW.

201 grid. The three-dimensional effects produce a smaller separated-flow region in the centerplane than that predicted in the two-dimensional calculations. Furthermore, the size of the separated-flow region and the pressure level in that region are in excellent agreement with the data for the finest mesh in the 3-D calculation. The time variation of the 3-D computation indicated a steady-state is established in approximately 4 ms, which is in agreement with the experiment. The extent of and the time to establish the separation are thus strongly influenced for this flow by three-dimensional effects.

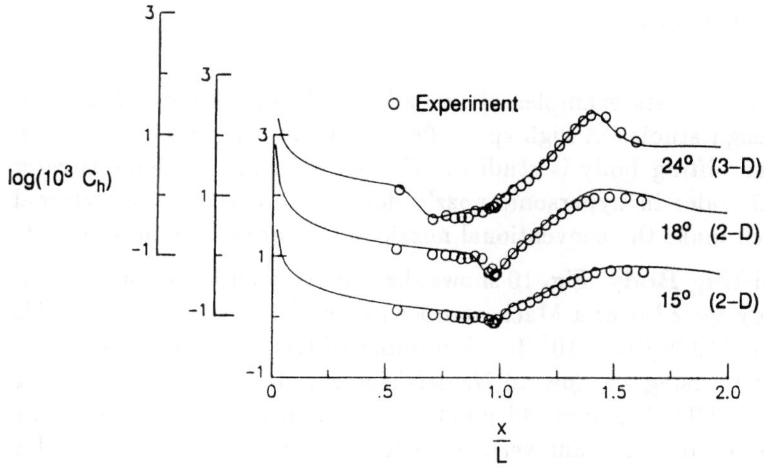

FIGURE 9. SUMMARY OF COMPUTATION AND EXPERIMENT FOR WEDGE ANGLES OF 15, 18, AND 24 DEGREES.

Fig. 9 shows a summary comparison of the heat transfer coefficient between computation and experiment for three different wedge angles. The 3-D solution in the centerplane is shown for the 24-degree wedge. For both the 15- and 18-degree cases, only 2-D solutions are shown since the extents of separation and hence the three-dimensional effects are significantly less. Excellent agreement of the computed heat transfer with experimental data for all three wedge angles is evident. It should be noted that the experimental data have been revised recently by Holden, who recomputed the free-stream conditions using the tunnel calibration rather than using pitot-tube measurements from probes within the shock layer emanating from the leading edge. The resulting change in the surface coefficient data improves the agreement of the predictions in the leading edge region.[26]

The results shown above reconcile several discrepancies between numerical computations and experiment for the 24-degree compression ramp presented in the literature and illustrates some of the difficulties encountered in validating CFD codes. The effects of grid refinement are substantial and modeling of the three-dimensional aspect of the experiment is critical. One difficulty in the process that is not uncommon is that the experiments are usually done prior to the computations (in this case by nearly two decades) and insufficient information exists of the detail required for code validation. A much better scenario is for the computations and experiments to be planned and conducted in a coordinated venture, much in the spirit of the present workshop.

Advanced Applications

This section presents examples of two advanced applications of CFD in analysis and design studies of high-speed flows. In the first application, flow over a hypersonic lifting body is studied. The second example demonstrates the use of CFD codes in hypersonic nozzle design subject to some external constraints which make the conventional nozzle design procedure impractical.

Hypersonic Lifting Body Fig. 10 shows the geometry and grid for a hypersonic lifting body tested over a Mach number range of 11 to 19 and Reynolds number range of 11,000/ft to 10^7/ft. A number of test conditions were simulated numerically using an upwind Navier-Stokes code.[29] A grid consisting of approximately 200,000 points (46 points axially and 65 x 65 points in the cross-plane) was used in the analysis. Numerical results are shown here for laminar flow at Mach 19.2 and Reynolds number 360,000/ft at zero angle of attack.

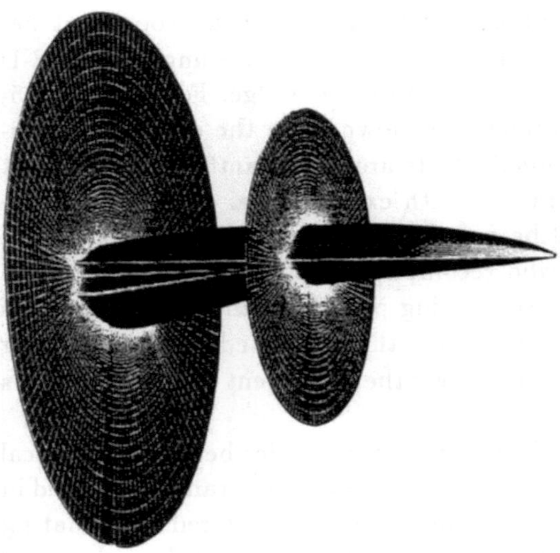

FIGURE 10. GEOMETRY AND GRID FOR HYPERSONIC LIFTING BODY.

Fig. 11 shows the particle traces around the lifting body. It is seen that for this configuration, the flow tends to pile up on the lower surface along the symmetry plane, which is not a very desirable feature as this low momentum fluid would enter the inlets of the propulsion system. Fig. 12 shows the comparison of predicted surface heating rates with the experimental results. The axial heating rates in the upper and lower symmetry planes and the cross-section heating rates presented indicate that the predicted heating rates generally compare well with the experimental data. More extensive results are presented in Ref. 29.

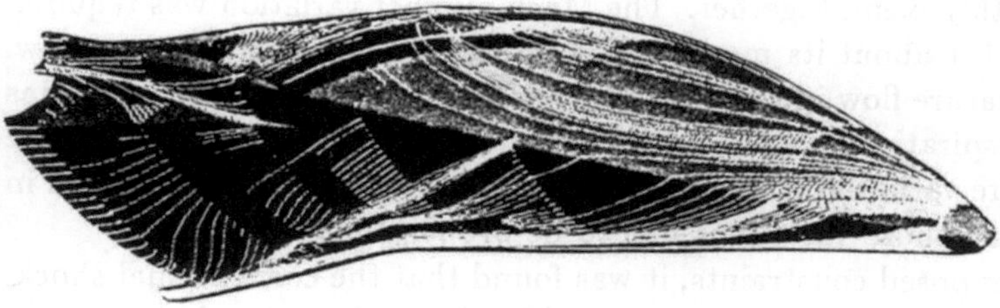

FIGURE 11. PARTICLE TRACES ON THE LIFTING BODY.

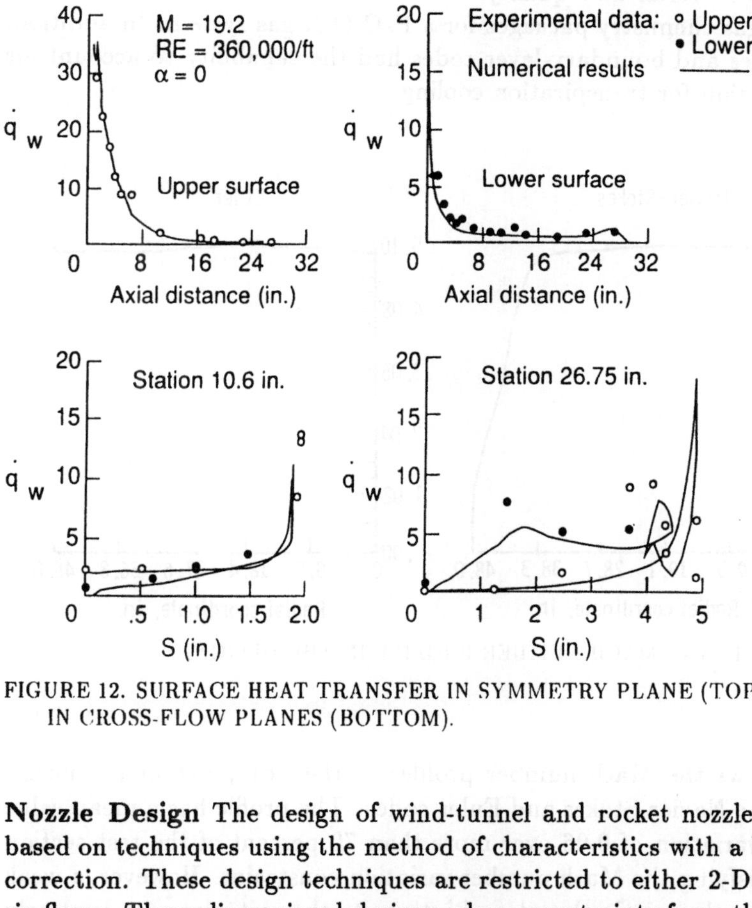

FIGURE 12. SURFACE HEAT TRANSFER IN SYMMETRY PLANE (TOP) AND IN CROSS-FLOW PLANES (BOTTOM).

Nozzle Design The design of wind-tunnel and rocket nozzles are generally based on techniques using the method of characteristics with a boundary-layer correction. These design techniques are restricted to either 2-D or axisymmetric flows. Three-dimensional designs whose constraints force the relaxation of the requirement of shockless flow will require more sophisticated CFD tools. The effort to design the nozzle contours for the NASA Langley 8 Foot High-Temperature Tunnel (HTT) provides an example of the use of advanced CFD codes in design projects. Under this effort, two nozzles for Mach 4 and 5 were designed such that they smoothly blend with the existing Mach 7 nozzle about 200 inches upstream of the test section. The specified constraints were: (1) the axial position and radius of the entrance to the subsonic region; (2) the throat axial location; and (3) the axial station where current and new nozzle walls must smoothly blend together. The Mach number variation was required to remain below 0.1 about its mean values across 60 percent of the core flow. The high temperature flow in the tunnel required the possibility of foreign gas injection for transpiration cooling in the nozzle throats. Furthermore, the large static temperature variation in the nozzles resulted in a significant variation in gas properties, and those variations had to be properly modeled.

Due to the imposed constraints, it was found that the conventional shockless nozzle design procedure was not applicable. A new iterative design procedure was developed[30] under this effort that coupled an Euler code, a method of characteristics code, and a boundary-layer code. A Navier-Stokes code was used to check the overall flow quality of the final design. All codes included consistent real gas chemistry packages for a H-C-O-N gas system. In addition, the Navier-Stokes and boundary-layer codes had the capability to account for foreign gas injection for transpiration cooling.

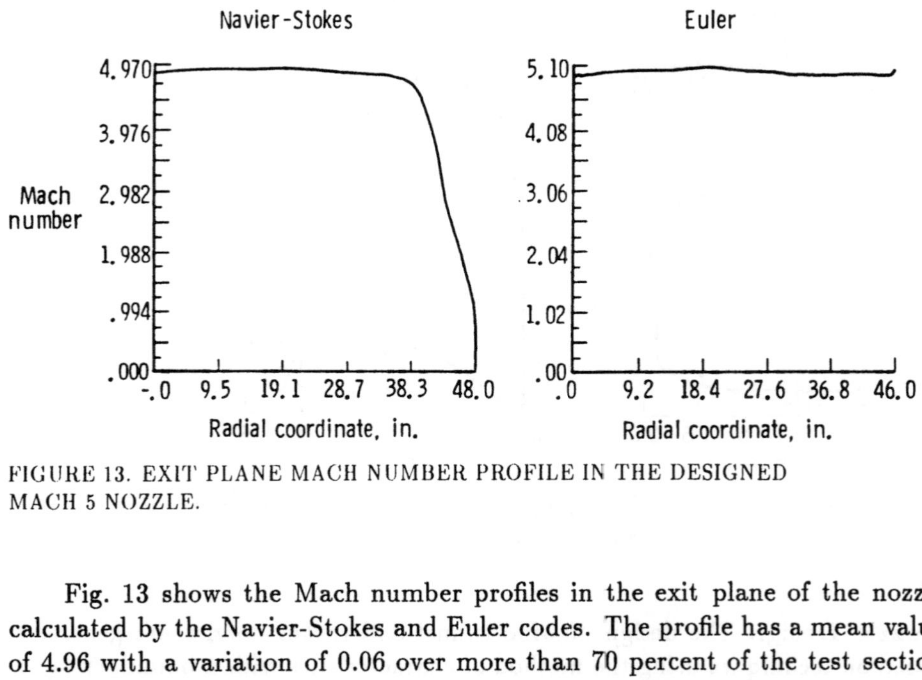

FIGURE 13. EXIT PLANE MACH NUMBER PROFILE IN THE DESIGNED MACH 5 NOZZLE.

Fig. 13 shows the Mach number profiles in the exit plane of the nozzle calculated by the Navier-Stokes and Euler codes. The profile has a mean value of 4.96 with a variation of 0.06 over more than 70 percent of the test section radius, thus satisfying the Mach number variation constraint. However, a weak shock forms near the nozzle throat and intersects the exit plane, as is clearly seen from the Mach number contours in Fig. 14. It appears that the weak shock cannot be avoided under the specified geometric constraints. The figures also illustrate the qualitative similarities in the flow solution obtained from the iterative design procedure and the Navier-Stokes code.

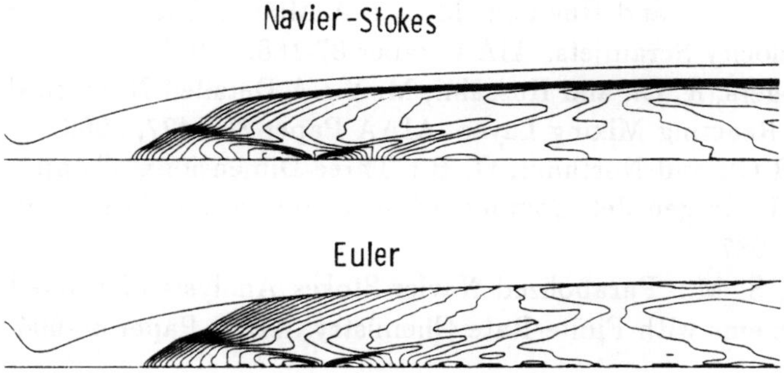

FIGURE 14. MACH NUMBER CONTOURS IN THE DESIGNED MACH 5 NOZZLE.

Concluding Remarks

The purpose of this paper has been to discuss CFD technology as it relates to the computation of flow fields associated with hypersonic airbreathing aircrafts. It appears at this point that CFD is the only available means for analyzing the flow over the entire flight trajectory because of the unavailability of experimental facilities at the high Mach number, high enthalpy conditions encountered. This is a new design environment and advances in algorithm robustness and speed, geometric flexibility, and inclusion of more complete flow physics are required in the CFD codes if they are to be widely used in the development of hypersonic airbreathing aircrafts. Although significant progress had been made over the past few years in the areas of transition, algorithms, and three-dimensional analysis and validation, for instance, there remain many important problems still to be solved. The most serious deficiencies lie in turbulence modeling, lack of complete transition prediction methods, and combustion modeling.

References

[1]Anderson, J. D., Jr.: A Survey of Modern Research in Hypersonic Aerodynamics. AIAA Paper 84-1578, 1984.

[2]Henry, J. R. ; and Anderson, G. Y.: Design Considerations for the Airframe Integrated Scramjet. NASA TM-X-2895, 1973.

[3]White, M. E.; Drummond, J. P.; and Kumar, A.: Evolution and Status of CFD Techniques for Scramjet Applications. J. of Propulsion and Power, Vol. 3, No. 5, Sept.-Oct. 1987, pp. 423-439.

[4]Kumar, A.; Graves, R. A., Jr.; Weilmuenster, K. J.; and Tiwari, S. N.: Laminar and Turbulent Flow Solutions with Radiation and Ablation Injection for Jovian Entry. Aerothermodynamics and Planetary Entry, Prog. in Astronautics and Aeronautics, Vol. 77, AIAA, New York, 1981, pp. 351-373.

[5] Dwoyer, D. L.; and Kumar, A.: Computational Analysis of Hypersonic Airbreathing Aircraft Flow Fields. AIAA Paper 87-0279, 1987.

[6]Kumar, A. Bushnell, D. M.; and Hussaini, M. Y.: A Mixing Augmentation Technique for Hypervelocity Scramjets. AIAA Paper 87-1882, 1987.
[7]Drummond, J. P.; Rogers, R. C.; and Hussaini, M. Y.: A Detailed Numerical Study of a Supersonic Reacting Mixing Layer. AIAA Paper 86-1427, 1986.
[8]Uneshi, K.; Rogers, R.C.; and Northam, G. B.: Three-Dimensional Computations of Transverse Hydrogen Jet Combustion in a Supersonic Airstream. AIAA Paper 87-0089, 1987.
[9]Sinha, N.; and Dash, S. M.: Parabolized Navier-Stokes Analysis of Ducted Turbulent Mixing Problems with Finite-Rate Chemistry. AIAA Paper 86-004, 1986.
[10]Chitsomboon, T.; Kumar, A.; Drummond, J. P.; and Tiwari, S. N.: Numerical Study of Supersonic Combustion Using a Finite-Rate Chemistry Model. AIAA Paper 86-0309, 1986.
[11]Schiff, L. B.; and Steger, J. L.: Numerical Simulation of Steady Supersonic Viscous Flow. AIAA Paper 79-0130, 1979.
[12]Kaul, U. K.; and Chaussee, D. S.: AFWAL Parabolized Navier-Stokes Code: 1983 AFWAL/NASA Merged Baseline Version. AFWAL-TR- 83-3118, May 1984.
[13]Gnoffo, P. A.: Hypersonic Flows Over Biconics Using a Variable- Effective Gamma Parabolized Navier-Stokes Code. AIAA Paper 83-1666, 1983.
[14]MacCormack, R. W.: The Effect of Viscosity in Hypervelocity Impact Cratering. AIAA Paper 69-354, 1969.
[15]Kumar, A.: Numerical Simulation of Scramjet Inlet Flow Fields. NASA TP-2517, May 1986.
[16]Beam, R.; and Warming, R. F.: An Implicit Factored Scheme for the Compressible Navier-Stokes Equations. AIAA J., Vol. 16, April 1978, pp. 393-402.
[17]Jameson, A.; Schmidt, W.; and Turkel, E.: Numerical Solutions of t he Euler Equations by Finite-Volume Methods Using Runge-Kutta Time-Stepping Schemes. AIAA Paper 81-1259, 1981.
[18]Vatsa, V. N.: Accurate Numerical Solutions for Transonic Viscous Flows Over Finite Wings. J. of Aircraft, Vol. 24, No. 6, June 1987, pp. 377-385.
[19]Thomas, J. L.; and Walters, R. W.: Upwind Relaxation Algorithms for the Navier-Stokes Equations. AIAA Paper 85-1501-CP, 1985.
[20]Chakrarthy, S. R.; Szema, K.-Y.; Goldberg, U.; Gorski, J. J.; and Osher, S.: Application of a New Class of High Accuracy TVD Schemes to the Navier-Stokes Equations. AIAA Paper 85-0165, 1985.
[21]Gnoffo, P. A.: Application of Program LAURA to Three-Dimensional AOTV Flow Fields. AIAA Paper 86-0565, 1986.
[22]Barber, T. J.; and Cox, G. B., Jr.: Hypersonic Vehicle Propulsion: A CFD Application Case Study. AIAA Paper 88-0475, 1988.
[23]Marvin, J. G. and Holst, T. L.: CFD Validation for Aerodynamic Flows-Challenge for the 90's. AIAA 90-2995 CP, 1990.
[24]Holden, M. S.; and Moselle, J. R.: Theoretical and Experimental Studies of the Shock Wave-Boundary Layer Interaction on Compression Surfaces in Hypersonic Flow. ARL 70-0002, January 1970.
[25]Hung, C. M.; and MacCormack, R. W.: Numerical Solutions of Supersonic

and Hypersonic Laminar Compression Corner Flows. AIAA J., Vol. 14, No. 4, April 1976, pp. 475-481.

[26]Rudy, D. H.; Thomas, J. L.; Kumar, A.; Gnoffo, P. A.; and Chakravarthy, S. R.: A Validation Study of Four Navier-Stokes Codes for High-Speed Flows. AIAA Paper 89-1838, June 1989.

[27]Putnam, L. E. : Investigation of Effects of Finite Span and Deflection Angle on Laminar Boundary-Layer Separation at Mach 10.03. NASA TN D-2833, May 1965.

[28]Ball, K. O. : Flap Span Effects on Boundary-Layer Separation. AIAA J., Vol. 9, No. 10, 1971, pp. 2080-1.

[29]Richardson, P. F.; Parlette, E. B.; Morrison, J. H.; Switzer, G. F. Dilley, A. D.; and Eppard, W. M.: Data Comparison Between a 3-D Navier-Stokes Analysis and Experiment for the McDonnell Douglas Lifting Body. Paper No. 52, Fourth National Aero-Space Symposium, Feb. 1988.

[30]Erlebacher, G.; Kumar, A.; Anderson, E. C.; Rogers, R. C.; Dwoyer, D. L.; Salas, M. D.; and Harris. J. E.: A Computational Design Procedure for Actively Cooled Hypersonic Wind-Tunnel Nozzles Subject to Wall Shape Constraints. Proceedings of CFD in Aerospace Design Workshop, University of Tennessee Space Institute, UTSI Pub. E02-4005-029-85, June 1985.

SIMILITUDE IN HYPERSONIC AERODYNAMICS

H. VIVIAND *

Office National d'Etudes et de Recherches Aérospatiales
ONERA - 29 Avenue Division Leclerc - 92320 Chatillon

ABSTRACT

This paper is devoted to a review and discussion of some aspects of similitude for hypersonic flows. Similitude is first defined in relation with fluid mechanics modelling, and a distinction is made between exact and approximate similitude for a given mathematical model. The general similitude for the full Navier-Stokes model, for a calorically perfect gas, is considered as a starting point and we deduce the approximate similitude which corresponds to the Mach number independence principle for large Mach numbers. These similitudes are shown to remain valid in the slip flow regime.

For inviscid hypersonic flows of a calorically perfect gas, we exhibit the similitudes which are associated with different models : Newtonian model, hypersonic small perturbation theory and Sytchev's model for slender bodies at large incidence. Similitude for viscous flows over slender bodies is considered next in the framework of hypersonic viscous interaction theory.

For hypersonic flows involving high temperature real gas effects, whether in thermodynamic equilibrium or not, exact similitude is not possible in general. We consider particular cases such as the thermally perfect gas and the simple dissociating diatomic gas.

I. INTRODUCTION

The very important role of similitude in classical fluid mechanics is well-known and is discussed in textbooks. The presentation which is given here is especially intended for computational fluid dynamics specialists, and it emphasizes the close connection which exists between mathematical modelling and similitude, whereas usual presentations rely more on the physics.

However, it is important to have a clear understanding of the physical basis of similitude, and for this, to have clear definitions of dimensional and dimensionless quantities involved in the model. This is why we start with a rather detailed definition of exact dimensional similitude, and then we distinguish it from extended mathematical similitude and from approximate similitude.

Similitude establishes a simple correspondence between different flow fields, based on the theoretical representation of these flows (which we shall refer to as mathematical model or model in short), without having to resort to the

---

* now at : PRINCIPIA - B.P. 122 - Sophia Antipolis - 06561 Valbonne Cédex

solution of this model. This correspondence then can be used to relate two real physical flow situations or to relate an infinity of solutions of the model.

Hypersonic flows constitute an interesting field of application of similitude because, on the one hand there exist many different hypersonic flow situations which can be described by a variety of models leading to different similitude laws and, on the other hand, real gas high temperature effects severely restrict exact similitude and provide incentive to study approximate similitudes.

Numerous studies of approximate models and of the corresponding similitudes for hypersonic flows have been carried out in the late fifties and early sixties. These studies have not lost their interest, despite the emergence and great success of computational fluid dynamics (CFD). However the aim of the present paper is not to make a review of all these pioneering works, but rather to give a CFD oriented view of similitude, with application to hypersonics. So only some representative references are given along the way, without any attempt to be complete.

Similitude in hypersonic flow has been discussed in depth by Hayes and Probstein [1], [2], especially for inviscid flows in the second edition of their book [2]. The book of Sedov [3], devoted to similitude in fluid mechanics, and an early ONERA report by Martinot-Lagarde [4] on dimensional analysis and similitude have provided us also with useful ground material.

II. DEFINITION OF SIMILITUDE

II.1 Dimensional similitude

Similitude establishes a simple connection between flow fields belonging to some subset of a given family of flows. In the following, by flow field (or flow, in short), we mean a unique physical flow configuration defined (in space and, possibly, in time) in terms of the values of physical dimensional variables in some chosen unit system.

This family of flows is essentially determined by the mathematical model chosen to describe the flows, i.e. which the flows are solutions of. The model includes the definition of the flow domain and of all the equations, relations, boundary conditions and initial conditions which (supposedly) make up a well-posed problem. Now it is important to recognize that the model involves physical or geometrical quantities which are constant for each flow, and which we shall call data (e.g., body length, freestream velocity, wall temperature, constants in the viscosity - temperature law, etc...) ; their values must be given to completely specify a flow field. Then the family, $\mathcal{F}$, is generated by varying all or some of the data within some ranges.

Let us note $\bar{D}_1$, $\bar{D}_2$, ... $\bar{D}_N$ all the independent physical data occurring in the model ; a set of values of these data can be represented by a point of the space R^N, and this point describes a domain $\mathcal{D}$ when the data are varied. There is a one to one correspondence between the flows of the family $\mathcal{F}$ and the points of the domain $\mathcal{D}$. Similitude amounts to an equivalence relation on the domain $\mathcal{D}$.

The data can be classified as follows :

• geometrical data defining the boundaries of the flow domain and in particular the obstacle surfaces ;

• physical data concerning the flow at the boundaries and entering into the boundary conditions ;

• physical data concerning the flow and entering into the initial conditions (if any) ;

• physical data concerning the intrinsic properties of the fluid; such as those occurring in thermodynamical equations, in constitutive relations for the stress tensor and the heat flux vector, in chemical rate equations, etc...

• physical data concerning the action of external systems (force, energy flux...) on the flow.

A flow field is described by a set of chosen dependent variables, for example :

(1) $\bar{\varphi} = (\bar{\rho}, \vec{\bar{U}}, \bar{h}, \bar{q}_1, \bar{q}_2, ...)$

where $\bar{\rho}$ is the density, $\vec{\bar{U}}$ the velocity, $\bar{h}$ the specific enthalpy and the $\bar{q}_i$ other possible variables (e.g. chemical species mass fractions). For each flow, these dependent variables are solutions of the mathematical model for given values of the data, and are functions of the independent spatial and, possibly, temporal variables $\bar{x}_j$ and $\bar{t}$. For all the flows of the family $\mathcal{F}$, these functions depend also on the data and can be written formally :

(2) $\bar{\varphi} = \bar{F}(\bar{x}_j, \bar{t}, \bar{D}_1, \bar{D}_2, ... \bar{D}_N)$

One particular flow of the family $\mathcal{F}$ is obtained by giving numerical values to the data. Here it is important to keep in mind the meaning of the notation. A symbol with a bar represents a physical quantity (i.e. a quantity having physical dimensions) and at the same time it represents the measure of this quantity in a chosen, fixed, unit systems ; this unit system need not be specified but it is the same for all the flows of $\mathcal{F}$: we shall refer to this system as the F.U.S. (Fixed Unit System).

Similitude arises from the fact that physical laws are invariant in a change of unit system (for coherent unit systems). Thus the mathematical model, hence the functions $\bar{F}$, are invariant in a change of unit system. So it is possible for each flow of $\mathcal{F}$, to use a particular unit system in which the fundamental units are related to the values of some of the data ; these systems will be referred to as the P.U.S. (Particular Unit Systems).

Four basic units are involved in general : length, time, mass and temperature. To fix ideas, we consider that we can select data, or combinations of data, which are the following :

a length $\bar{L}_*$, a velocity $\bar{U}_*$, a density $\bar{\rho}_*$ and a temperature $\bar{T}_*$.

These data will be called basic reference quantities ; they allow to define the fundamental units of the P.U.S. , through their measures in the F.U.S., as follows :

• unit of length $= \bar{L}_*$

• unit of time $= \bar{t}_* = \bar{L}_*/\bar{U}_*$,

• unit of mass $= \bar{M}_* = \bar{\rho}_* \bar{L}_*^3$,

• unit of temperature $= \bar{T}_*$

Then secondary units for all other quantities or variables follow uniquely from these basic reference quantities. For any quantity Ψ we note Ψ_* the measure in the F.U.S. of the corresponding P.U.S. unit ; it is of the form :

$\bar{\Psi}_* = \bar{L}_*^a \, \bar{U}_*^b \, \bar{\rho}_*^c \, \bar{T}_*^d$

where a, b, c, d are numbers. For example, we have :

- unit of pressure or stress : $\bar{p}_* = \bar{\rho}_* \, \bar{U}_*^2$
- unit of specific energy or enthalpy $\bar{h}_* = \bar{U}_*^2$
- unit of heat flux density : $\bar{q}_* = \bar{\rho}_* \, \bar{U}_*^3$
- unit of viscosity coefficient : $\bar{\mu}_* = \bar{\rho}_* \, \bar{U}_* \, \bar{L}_*$
- unit of thermal conductivity : $\bar{k}_* = \bar{\rho}_* \, \bar{U}_*^3 \, \bar{L}_* \, / \, \bar{T}_*$

Of course, a unique choice of data as basic reference quantities is made for all the flows of the family; the P.U.S. changes for each flow according to the values of these data. The variables and quantities now measured in the P.U.S. associated with each flow will be called dimensionless variables or quantities and will be written without a bar : $\Psi = \bar{\Psi} \, / \, \bar{\Psi}_*$

The dimensionless dependent variables then write :

(3) $\quad \varphi = (\rho, \, \vec{U}, \, h, \, q_1, \, q_2, ...)$

and the formal expression of the flows (2) is unchanged, due to the invariance of physical laws with respect to unit systems :

(4) $\quad \varphi = \bar{F} \, (x_j, \, t, \, D_1, \, D_2, \, ...D_N)$

where $\bar{F}$ is the same function as in Eq. (2).

For convenience, with a proper choice of independent data, we can always assume that the four basic reference quantities are the first four data :

$\bar{D}_1 = \bar{L}_*, \, \bar{D}_2 = \bar{U}_*, \, \bar{D}_3 = \bar{\rho}*, \, \bar{D}_4 = \bar{T}_*$

In the P.U.S., these same dimensionless data become equal to unity $D_i = 1, \; i = 1$ to 4, and the expression (4) can be written in the new form :

(5) $\quad \varphi = \Phi \, (x_j, \, t, \, P_1, \, P_2, \, ...P_{N-4})$

where the remaining dimensionless data D_i, $i = 5$ to N , are noted P_{i-4}, and are of the following form :

$$P_j = D_{j+4} = \bar{D}_{j+4} / \bar{D}_{j+4*}, \quad j = 1 \text{ to } N-4 \tag{6}$$

$$\bar{D}_{j+4*} = \bar{L}_*^{aj} \, \bar{U}_*^{bj} \, \bar{\rho}_*^{cj} \, \bar{T}_*^{dj}$$

In all that follows, the P_j are called parameters ; they are the similitude parameters. The passage from Eq. (2) to Eq. (5) amounts to the application of Vaschy's (or Pi) theorem.

The definition of similitude follows immediately. Different flows of the family $\mathcal{F}$, i.e. flows with different sets of values of data, are similar if these sets of data correspond to fixed values of the parameter P_j. The relation existing between similar flows is obtained by rewriting Eq. (5) in terms of physical quantities (but keeping the parameters fixed) :

(7) $\quad \bar{\varphi} = \bar{\varphi}_* \, \Phi \, (\bar{x}_j / \bar{L}_*, \;\; \bar{U}_* \bar{t} / \bar{L}_*, \;\; P_1, \, P_2, \, ...P_{N-4})$

where $\bar{\varphi}_* = \bar{\rho}_*, \, \bar{U}_*, \, \bar{h}_*$, or $\bar{q}_{\ell *}$

Similar flows are obtained from Eq. (7) by varying the data (within some limits) in all possible ways which keep all the parameter values fixed. The knowledge of one flow leads immediately, through Eq. (7), to the knowledge of all the flows similar to that one.

II.2 Extended mathematical similitude

It may happen, depending on the model, that a reduction of the number of parameters is possible by using new dependent and independent variables resulting from affine transformations which are functions of the parameters. Let us note :

$\varphi = (\varphi_1, \varphi_2, ...)$ and $\varphi_j = \Phi_j(x_i,\ t,\ P_1,\ P_2, ...P_{N-4})$

The transformation is of the form :

(8) $\quad x_i = \tilde{x_i}\ \tau_{xi}(P_k),\ t = \tilde{t}\ \tau_t(P_k),\ \varphi_j = \tilde{\varphi_j}\ \tau_{\varphi j}\ (P_k)$

where the factors τ are functions of the parameters P_k (some of these factors may be equal to unity).

The expression (5) transforms into :

(9) $\quad \tilde{\varphi_j} = \tilde{\Phi_j}\ (\tilde{x_i},\ \tilde{t},\ Q_1,\ Q_2,\ ...Q_M)$

where the new parameter Q_m, $m = 1$ to M, are functions of the P_k :

(10) $\quad Q_m = Q_m\ (P_1,\ P_2,\ ...P_{N-4}),\ m = 1$ to M

A new similitude is obtained, which is more general than the original one if $M < N - 4$ since then there are less constraints on the variations of data. This property of the transformation (8), namely $M < N - 4$, cannot be verified on the solution itself because the Φ_j are not known explicitly in general, but it can easily be verified from the mathematical model, by carrying out the transformation (8) on the complete dimensionless model.

For completeness, we just mention the case when the number of independent variables (x_i, t) can be reduced through combinations involving them and also possibly the dependent variables. This leads to self-similar flows, where similarity properties apply within each flow, and this occurs when the data are not sufficient to provide all the basic reference quantities.

II.3 Approximate similitude

If one of the parameters, say P_1, has only a weak influence on the dimensionless flow, i.e. on the function Φ (Eq. (5)), then we can consider an approximate similitude in which the parameter P_1 is allowed to vary in some range where its influence remains weak.

The data $\bar{D_1}$ to $\bar{D_N}$ being independant, we can choose to vary only the datum $\bar{D_5}$ which corresponds to P_1 (Eq. (6)), thus keeping all the parameters fixed except P_1. Then, by hypothesis, we observe no appreciable effect on the dimensionless flow. This means that the datum $\bar{D_5}$ plays a weak role in the problem and can be varied (within some limits) independently of the other data, thus enlarging the class of similar flows. Of course, such an approximate similitude can be exhibited only if the datum which has a weak influence, here $\bar{D_5}$, has not been chosen as one of the basic reference quantities.

In general such a behaviour is obtained for very small or very large values of the parameter P_1, the dimensionless flow having a limit as P_1 tends to zero or to infinity. Whereas the exact similitude does not depend on the choice of the basic reference quantities, the limit behavior of the flow can be exhibited only if the proper dimensionless variables are used, i.e. if the order of magnitude of these variables is constant as the parameter approaches its limit value.

III. GENERAL SIMILITUDE FOR CALORICALLY PERFECT GAS FLOWS

III.1 Navier-Stokes similitude

To study the dimensional similitude associated with a given mathematical model, the first step is to explicitly write the complete mathematical model in the fixed unit system (F.U.S.) so as to identify all the independent data $\bar{D}_i$ of the model. Then the basic reference quantities are chosen among the data, and the particular unit system (P.U.S.) is thus fixed for each flow. The dimensionless form of the model (i.e. in the P.U.S.) is then written and all the similitude parameters can be identified.

As long as the basic reference quantities are not specified, the mathematical model has strictly the same form in the F.U.S. and in the P.U.S., the only difference being in the use of different symbols for the quantities in the F.U.S. and the P.U.S. (here with or without a bar). So, for brevity, we directly write the model in dimensionless form, but knowing that the basic reference quantities have not been chosen yet.

The model considered is based on the steady Navier-Stokes equations for a calorically perfect gas. The basic dependent variables which we use are the density ρ, the velocity vector $\vec{U}$, and the specific enthalpy h. The dimensionless model writes, in conservative form :

(1) $\quad div\ (\rho\ \vec{U}) = 0$

(2) $\quad div\ (\rho\ \vec{U} \otimes \vec{U} + p\ \vec{I}) = div\ \vec{\tau}$

(3) $\quad div\ (\rho\ \vec{U}\ H) = div\ (\vec{U}.\vec{\tau} - \vec{q})$

(4) $\quad H = h + \vec{U}^2/2$

(5) $\quad p = \rho\ \mathcal{R}\ T$

(6) $\quad h = C_p\ T$

(7) $\quad \vec{\tau} = \mu \left[grad\vec{U} + t\ (grad\vec{U}) - \frac{2}{3}\ div\vec{U}\ \vec{I} \right]$

(8) $\quad \vec{q} = -k\ \text{grad}T$

where p is the pressure, $\vec{I}$ the unit tensor, $\vec{\tau}$ the viscous stress tensor, $\vec{q}$ the heat flux vector, H the total enthalpy and T the absolute temperature.

For the viscosity, we consider the following dependency on temperature, which combines Sutherland's law and a power law :

(9) $\quad \mu = D_\mu\ T^{\omega+1}\ (T + T_S)^{-1}$

For $w = \frac{1}{2}$, one recovers Sutherland's law, and for $T_S = 0$, a simple power law with exponent ω.

The thermal conductivity is taken proportional to the viscosity :

(10) $k = D_k \, \mu$

These equations must be completed by boundary conditions. We consider unbounded flows past fixed obstacles belonging to some family given by the equation :

(11) $B(\vec{x}\ ,\ G_1\ ,\ G_2\ ,...)\ =\ 0$

where G_i are geometrical parameters. We assume that the following conditions ensure a well-posed problem :

– uniform flow upstream (at a finite distance, $x < x_0$, assuming supersonic upstream flow) :

(12) $$\begin{cases} a)\ \vec{U} = U_\infty \vec{i}, \text{ where } \vec{i} = \text{unit vector of x-axis} \\ b)\ \rho = \rho_\infty \\ c)\ h = h_\infty \end{cases}$$

– no-slip condition on the body $B = 0$:

(13) $\vec{U} = 0$

– thermal condition on the body $B = 0$; either the wall is at a fixed temperature :

(14) $T = T_w$,

or it is adiabatic :

(15) $gradB \cdot gradT = 0$

We recall that the model in physical variables (i.e. in the F.U.S.) can be recovered just by putting a bar over the symbols since the P.U.S. has been left undetermined ; so, the data are easily identified :

• $\bar{\mathcal{R}}$, the gas constant $(\mathcal{R} = \bar{\mathcal{R}}\, \bar{T}_* / \bar{U}_*^2)$

• $\bar{C}_p$, the specific heat at constant pressure $(C_p = \bar{C}_p\, \bar{T}_* \,/\, \bar{U}_*^2)$

• $\bar{D}_\mu$, $(D_\mu = \bar{D}\mu\, \bar{T}_*^{\omega} /(\bar{\rho}_*\, \bar{U}_*\, \bar{L}_*))$

• $\bar{T}_S$, $(T_S = \bar{T}_S / \bar{T}_*)$

• $\bar{\omega}$, $(\omega = \bar{\omega}$, pure number)

• $\bar{D}_k$ $(D_k = \bar{D}_k\, \bar{T}_* \,/\, \bar{U}_*^2)$

• $\bar{U}_\infty$ $(U_\infty = \bar{U}_\infty / \bar{U}_*)$

• $\bar{\rho}_\infty$ $(\rho_\infty = \bar{\rho}_\infty / \bar{\rho}_*)$

• $\bar{h}_\infty$ $(h_\infty = \bar{h}_\infty / \bar{U}_*^2)$

• $\bar{T}_w$ $(T_w = \bar{T}_w \,/\, \bar{T}_*)$, in the case of condition (14)

• $\bar{G}_1$, $\bar{G}_2$..., geometrical data ; among them, we choose a characteristic length $\bar{L}$, say $\bar{G}_1 = \bar{L}$

The basic reference quantities are then chosen as follows :

(16) $\bar{L}_* = \bar{G}_1 = \bar{L}\,,\ \bar{U}_* = \bar{U}_\infty\,,\ \bar{\rho}_* = \bar{\rho}_\infty\,,\ \bar{T}_* = \bar{U}_\infty^2 \,/\, \bar{\mathcal{R}}$

hence we have : $G_1 = 1,\ U_\infty = 1,\ \rho_\infty = 1,\ \mathcal{R} = 1$

Other choices are possible, but the exact similitude does not depend on this choice. The **similitude parameters** are then found to be :

(17) • $C_p = \bar{C}_p \,/\, \bar{R} = \gamma \,/\, (\gamma - 1)$

where γ is the usual specific heat ratio.

(18) • $D_\mu = \bar{D}_\mu (\bar{U}_\infty^2/\mathcal{R})^\omega \,/\, (\bar{\rho}_\infty \bar{U}_\infty \bar{L})$

(19) • $T_S = \bar{T}_S \bar{\mathcal{R}} \,/\, \bar{U}_\infty^2$

• ω

Instead of D_μ we can use :

(20) $R_e = (1 + T_S) \,/\, D_\mu$

where R_e is the Reynolds number based on freestream velocity and density and on the viscosity at the temperature $\bar{T}_*$:

(21) $R_e = \bar{\rho}_\infty \bar{U}_\infty \bar{L} \,/\, \bar{\mu}\,(\bar{T}_*)$

(22) • $D_k = \bar{D}_k \,/\, \bar{\mathcal{R}} = \bar{k} \,/\, (\bar{\mu}\,\bar{\mathcal{R}}) = \gamma \,/\, ((\gamma - 1)\,\mathcal{P}_r)$

where $\mathcal{P}_r$ is the Prandtl number.

(23) • $h_\infty = \bar{h}_\infty \,/\, \bar{U}_\infty^2 = ((\gamma - 1)\, M_\infty^2)^{-1}$

where $M_\infty = \bar{U}_\infty \,/\, ((\gamma - 1)\, \bar{h}_\infty)^{\frac{1}{2}}$ is the free-stream Mach number.

(24) • $T_w = \bar{T}_w \bar{\mathcal{R}} \,/\, \bar{U}_\infty^2 = (\bar{T}_w \,/\, \bar{T}_\infty)\, h_\infty\, (\gamma - 1) \,/\, \gamma$

(25) • $G_i = \bar{G}_i \,/\, \bar{L}\,,\ (i > 1)$

With this choice of P.U.S., Eq. (5) and the boundary conditions (12-a, 12-b) simplify as :

(26) $p = \rho\, T$

(27) $\vec{U} = \vec{i}$ for $x < x_0$

(28) $\rho = 1$ for $x < x_0$

The equations (1) to (4) involve no parameter and so cannot be affected by the choice of P.U.S. The equations (6) to (11) and the conditions (12-c) and (13) to (15) remain unchanged, but the parameters involved are now related to the data by the above relations (note that the conditions on the body involve the parameters G_i if such parameters are present). In particular, we can write :

(29) $h = T\,\gamma \,/\, (\gamma - 1)$

One often makes use of the Reynolds number based on freestream viscosity, i.e. :

(30) $R_\infty = \bar{\rho}_\infty \bar{U}_\infty \bar{L} \,/\, \bar{\mu}\,(\bar{T}_\infty)$

which is related to R_e (Eq. 20) by :

(31) $R_\infty = R_e\, (\gamma\, M_\infty^2)^{\omega+1}\, [T_S + (\gamma\, M_\infty^2)^{-1}] \,/\, (T_S + 1)$

Similar flows can then be obtained, for example, by free and independent variations of the following data :

$\bar{\mathcal{R}}$, $\bar{D}_\mu$, $\bar{U}_\infty$ and $\bar{\rho}_\infty$

while the other data are constrained to vary as follows (using the sign $\sim$ to express proportionality) :

$\bar{C}_p \sim \bar{\mathcal{R}}$, $\bar{T}_S \sim \bar{U}_\infty^2 \,/\, \bar{\mathcal{R}}$, $\omega = Cst$, $\bar{D}_k \sim \bar{\mathcal{R}}$, $\bar{h}_\infty \sim \bar{U}_\infty^2$ (i.e. $\bar{T}_\infty \sim \bar{U}_\infty^2 \,/\bar{\mathcal{R}}$) , $\bar{T}_w \sim \bar{U}_\infty^2 \,/\, \bar{\mathcal{R}}$, $\bar{L} \sim \bar{D}_\mu \, U_\infty^{2\bar{\omega}-1} \,/\, (\bar{\rho}_\infty \, \bar{\mathcal{R}}^\omega)$.

It is of interest to determine similar flows of a given gas, i.e. to fix the data $\bar{\mathcal{R}}$, $\bar{C}_p$, $\bar{D}_\mu$, $\bar{T}_S$, ω and $\bar{D}_k$. Then we find that similitude requires to fix also $\bar{U}_\infty$, $\bar{h}_\infty$, and$\bar{T}_W$ and that we only vary $\bar{\rho}_\infty$ and $\bar{L}$ with $\bar{L}\,\bar{\rho}_\infty = Cst$. In the case of a power law, due to the fact that $\bar{T}_S = 0$ so that T_S disppears as a similitude parameter, it is possible to vary $\bar{U}_\infty$ and $\bar{\rho}_\infty$ independently while imposing :

$\bar{h}_\infty \sim \bar{U}_\infty^2$, $\bar{T}_w \sim \bar{U}_\infty^2$, $\bar{L} \sim \bar{U}_\infty^{2\omega-1}/\bar{\rho}_\infty$.

We see that the similitude depends strongly on the choice of the viscosity law.

In the case of inviscid flows described by the Euler equations ($\bar{\mu} = \bar{k} = 0$), the data $\bar{D}_\mu, \bar{T}_S$, ω, $\bar{D}_k$ and $\bar{T}_w$ disappear and the similitude parameters reduce to C_p and h_∞ (i.e. γ and M_∞) and the geometrical parameters. Similar flows are generated, for example, by free variations of $\bar{\mathcal{R}}$, $\bar{U}_\infty$, $\bar{\rho}_\infty$, and $\bar{L}$, while for the other data we have : $\bar{C}_p \sim \bar{\mathcal{R}}$ and $\bar{h}_\infty \sim \bar{U}_\infty^2$.

III.2 Extension to slip flow regime

When the gas density is very low, the local molecular mean free-path $\bar{\ell}$ may no longer be very small compared to a local characteristic length for the variations of macroscopic quantities, in which case the Navier-Stokes model becomes invalid and must be replaced by the Boltzmann equation.

However, when rarefaction effects are just incipient, they are first felt at solid boundaries in the form of velocity and temperature jumps between gas and solid surface. The Navier-Stokes equations (1) to (8) remain valid, but the conditions (13) and (14) on a solid surface must be replaced by jump relations provided by the kinetic theory of gases and called slip conditions. We examine the consequences of these modifications on the general similitude.

Define : $\vec{n} = gradB \,/\, |\, gradB \,|$, the unit normal vector on the body $B = 0$, and let $\vec{U} = (\vec{U}\cdot\vec{n})\,\vec{n} + \vec{V}$, and $\vec{V}$ is the tangential velocity along a normal to the body.

The slip conditions on $B = 0$, in dimensionless variables, write :

(32) $\quad \vec{U}.\vec{n} = 0, \;\; \vec{V} = ((2-\sigma)\,/\,\sigma)\,\ell\,(\partial\vec{V}\,/\,\partial n)$

(33) $\quad T - T_w = ((2-\alpha)\,/\,\alpha)\,(2\gamma/(\gamma+1))\,(1/\mathcal{P}_r)\,\ell\,(\partial T/\partial n)$

The molecular mean free-path ℓ is given by elementary kinetic theory in terms of macroscopic flow quantities by :

(34) $\quad \ell = (\pi/2)^{\frac{1}{2}}\,(\mu/\rho)\,[h(\gamma-1)/\gamma]^{-\frac{1}{2}}$

The constants σ and α are respectively the momentum and the thermal accomodation coefficients ; they are dimensionless data and therefore also new parameters. We see that these are the only new data and new parameters introduced by the slip conditions. They must have constant values for all similar flows, and the general similitude discussed in § III.1 applies without other changes.

It is usual to characterize the rarefied flow regimes by the value of the Knudsen number, which is the ratio of the molecular mean free-path to a characteristic macroscopic length. A global Knudsen number can be defined, based on the freestream conditions and the reference length $\bar{L}$:

$$(35) \qquad K_{n\infty} = \bar{\ell}_\infty \,/\, \bar{L} = \ell_\infty = (\pi/2)^{\frac{1}{2}}\ \mu_\infty\ [(h_\infty(\gamma-1)/\gamma]^{-\frac{1}{2}}$$

It can be expressed in terms of the upstream Mach number M_∞ and Reynolds number R_∞ defined by Eq. (30) :

$$(36) \qquad K_{n\infty} = (\pi\gamma/2)^{\frac{1}{2}}\ M_\infty / R_\infty$$

which confirms that it is not a new independent parameter.

IV HYPERSONIC FLOWS OF CALORICALLY PERFECT GASES

IV.1 Hypersonic limit and Mach number independence principle

We consider here the general similitude of §III in the case when the parameter h_∞ is very small (i.e. when the upstream Mach number M_∞ is very large) in order to extend the class of similar flows through an approximate similitude, as discussed in §II.3.

Such an approximate similitude results from the application of the limiting process $h_\infty \to 0$, keeping all other parameters fixed, to the proper dimensionless form of the mathematical model, i.e. using the dimensionless variables which remain of constant order of magnitude in this process. The choise of basic reference quantities made in §III.1. (Eq. (16)) leads to the proper dimensionless variables for this hypersonic limit (strong shock limit, in which all geometrical parameters remain fixed). The limit model is thus obtained from the general model presented in §III.1 simply by replacing the boundary condition (12.c) by :

$$(1) \qquad h_\infty = 0 \ \text{ for } x < x_0$$

For small enough values of the parameter h_∞, the dimensionless flow solution of the full model of §III.1 will remain close to the solution of this limit model wich is independent of h_∞, and we obtain an approximate extended similitude by letting the parameter h_∞ (or equivalently the Mach number M_∞), free to vary below some upper bound (or above some lower bound for the Mach number). This bound depends of course on the body shape ; also, the limit solution may not be uniformly valid over the entire flow field and the approximate similitude may apply only to the front part of the flow.

This is the well known Mach number independence principle (see Ref. [2] for a detailed discussion), usually considered for inviscid flows, but which holds also for viscous flows on condition to keep all the proper parameters fixed. In particular, the Reynolds number which enters into the limit model and which must be kept fixed is R_e, and not R_∞ ; from Eq. (31) of §III, we see that R_∞ tends to infinity in the hypersonic limit.

For similar flows of a fixed gas, $\bar{U}_\infty$, $\bar{T}_w$, and $(\bar{\rho}_\infty \bar{L})$ must be maintained fixed, as we have seen in §III.1.; but now h_∞ hence $\bar{h}_\infty$ (or $\bar{T}_\infty$) can vary freely as long as $\bar{h}_\infty \ll \bar{U}^2_\infty$.

With a power law viscosity ($\bar{T}_S = 0$), not only $\bar{h}_\infty$ but also $\bar{U}_\infty$ can vary freely (but again with $\bar{h}_\infty \ll \bar{U}^2_\infty$), and as we have seen in §III.1., $\bar{T}_w$ must vary as $\bar{U}^2_\infty$, and $(\bar{\rho}_\infty\ \bar{L})$ as $\bar{U}^{2\omega-1}_\infty$.

When slip flow conditions are taken into account in the general Navier-Stokes model (§III.2), the global Knudsen number $K_{n\infty}$ (Eq.(36) of §III.2 is no more a constant in this hypersonic approximate similitude, but it varies with M_∞ according to the following relations :

$$(2) \qquad Kn_\infty = \left(\frac{\pi}{2}\right)^{1/2} \frac{1}{R_e} (\gamma M^2_\infty)^{1/2-\omega} \frac{1+T_S}{(1+\gamma M^2_\infty\ T_S)}$$

IV.2 Inviscid flows

IV.2.1 Newtonian model

In the case of inviscid flows of calorically perfect gases, described by the Euler equations, the similitude parameters reduce to γ, h_∞ (or M_∞) and the geometrical parameters.

Here we depart slightly from assumption of a perfect gas made in §IV, and we allow a discontinuity in the specific heat ratio across the front shock. The gas still behaves as a perfect gas, but with two different values of γ, say γ_1 and γ_2 respectively in the freestream and in the perturbed flow. Such a discontinuity is a crude way to represent equilibrium real gas effects due to the high temperature behind the front shock which leads to a decrease in the value of γ.

The usual Newtonian limit is obtained as γ_2 tends to 1 and M_∞ to infinity simultaneously. Here we consider a more general Newtonian limit, as studied by Guiraud [5, 6], in wich γ_2 tends to 1 at fixed M_∞. Of course the upstream Mach number has to be large for γ_2 to be close to 1, but we keep it fixed. The density ρ_s, the normal velocity U_{ns} and the specific enthalpy h_s just behind the front shock are given, in the dimensionless variables of the P.U.S. defined in § III.1, by relations :

$$(3) \qquad \frac{1}{\rho_s} = \frac{U_{ns}}{sin\beta} = \frac{\gamma_2}{\gamma_2+1}\left[1+\frac{1}{\gamma_1\ M^2_{n\infty}}\right]\frac{\Delta}{1+(1-\Delta)^{1/2}}$$

with

$$\Delta = \left(\frac{\gamma_2^2-1}{\gamma_2^2}\ \left(1+\frac{2}{(\gamma_1-1)\ M^2_{n\infty}}\right)\ \left(1+\frac{1}{\gamma_1\ M^2_{n\infty}}\right)\right)^{-2}$$

$$(4) \qquad h_s = \frac{1}{2}\ sin^2\beta\ \left(1-\frac{1}{\rho_s^2}\right) + \frac{1}{(\gamma_1-1)\ M^2_{n\infty}}$$

where β is the angle between the upstream flow velocity and the shock surface, and $M_{n\infty} = M_\infty sin\beta$ is the normal Mach number. When γ_2 tends to 1, ρ_s tends to infinity as $\frac{1}{\gamma_2-1}$ and the perturbed flow is confined to a thin layer attached to the wall. To describe the Newtonian flow model, we use a system of orthogonal curvilinear coordinates x_1, x_2, x_3 such that $x_3 = 0$ is the surface of the body ; we note h_1, h_2, h_3 the corresponding metric coefficients, given functions of the coordinates, and U_1, U_2, U_3 the corresponding velocity components. We define the stretched coordinate Z and the transformed density $\tilde{\rho}$ and normal velocity W, by :

$$(5) \qquad x_3 = (\gamma_2 - 1)\,\frac{Z}{2},\; U_3 = (\gamma_2 - 1)\,\frac{W}{2},\; \tilde{\rho} = (\gamma_2 - 1)\,\frac{\rho}{2}$$

The Newtonian model is obtained from the Euler equations as a result of the limiting process $\gamma_2 \to 1$, in these new dimensionless variables, and it can be written as follows :

Continuity equation :

$$(6) \qquad \frac{\partial(h_2 h_3 \tilde{\rho} U_1)}{\partial x_1} + \frac{\partial(h_3 h_1 \tilde{\rho} U_2)}{\partial x_2} + \frac{\partial(h_1 h_2 \tilde{\rho} W)}{\partial Z} = 0$$

Tangential momentum equations :

$$(7) \qquad \begin{cases} a) \quad ((\vec{U}.\text{grad})\ \vec{V}).\vec{e}_1 = 0 \\ b) \quad ((\vec{U}.\text{grad})\ \vec{V}).\vec{e}_2 = 0 \end{cases}$$

where $\vec{V} = U_1\ \vec{e}_1 + U_2\ \vec{e}_2$, and $\vec{e}_1, \vec{e}_2, \vec{e}_3$ are the unit vectors of the local basis ; the material derivative writes :

$$\vec{U}.\text{grad} = \left(\frac{U_1}{h_1}\right)\frac{\partial}{\partial x_1} + \left(\frac{U_2}{h_2}\right)\frac{\partial}{\partial x_2} + \left(\frac{W}{h_3}\right)\frac{\partial}{\partial Z}$$

A consequence of Eqs. (7) is :

$$(8) \qquad \vec{U}.\text{grad}\,(\vec{V}^2) = 0$$

Normal momentum equation :

$$(9) \qquad \frac{1}{\tilde{\rho}}\frac{\partial p}{\partial Z} = \frac{U_1^2}{h_1}\frac{\partial h_1}{\partial x_3} + \frac{U_2^2}{h_2}\frac{\partial h_2}{\partial x_3}$$

Energy equation :

$$(10) \qquad \vec{U}.\ \text{grad}\ h = 0$$

State equation :

$$(11) \qquad p = 2\tilde{\rho}\ h$$

In these equations all metric terms are taken on the body surface, i.e. at $Z = 0$. The boundary conditions are :

$$(12) \qquad W = 0, \text{on the body } Z = 0$$

and on the downstream side of the front shock, at $Z = Z_s\ (x_1, x_2)$, (an unknown of the problem) :

$$(13)\quad \begin{cases} a)\ \tilde{\rho} = \tilde{\rho}_s = \left[1+\dfrac{1}{\gamma_1\, M_{n\infty}^2}\right]\left[1+\dfrac{2}{(\gamma_1-1)\, M_{n\infty}^2}\right]^{-1} \\ b)\ W = W_s = \dfrac{sin\beta}{\tilde{\rho}_s} \\ c)\ h = h_s = \dfrac{1}{2} sin^2\beta + ((\gamma_1-1)\, M_{n\infty}^2)^{-1} \end{cases}$$

In these shock relations, the shock normal can be taken equal to the body normal at the point having same x_1, x_2 coordinates.

The parameters, besides the geometrical ones, are only γ_1 and M_∞ ; γ_2 does not appear. So the pressure and the enthalpy are independent of γ_2, but the density is inversely proportional to $(\gamma_2 - 1)$. The only constraints on the data variations for similar flows are :

$\bar{C}_{p_1}/\bar{\mathcal{R}}_1 = Cst.$ (where the subscript 1 refers to the conditions upstream of the shock), and $\bar{h}_\infty/\bar{U}_\infty^2 =$ Cst.

If we let M_∞ tend to infinity, the conditions at the shock simplify into :

(14) $\quad a)\ \tilde{\rho} = 1, \quad b)\ W = \sin\beta, \quad c)\ h = \frac{1}{2} sin^2\beta$

so that γ_1 and M_∞ disappear from the limit dimensionless problem which has no similitude parameter : all the flows obeying the more general model defined by Eqs. (5) to (13) become approximately similar for large enough Mach numbers.

IV.2.2 Inviscid hypersonic flows past slender bodies

IV.2.2.1 Approximate model for the transversal flow

The 3D Euler equations for steady inviscid flow can be approximated, in the case of the hypersonic flow past a slender body at arbitrary incidence, by a model which is formally identical to the unsteady 2D Euler equations. This is the so-called equivalence principle which we briefly recall here (Refs. [2], [7]).

We consider a family of slender bodies which is described by the following equation, in a cartesian coordinate system attached to the body :

(15) $\quad B\,(x,\ \tilde{y}, \tilde{z}) = 0$ where $x = \bar{x}/\bar{L},\ \tilde{y} = \bar{y}/\bar{d},\ \tilde{z} = \bar{z}/\bar{d}$

where $\bar{L}$ and $\bar{d}$ are characteristic longitudinal and transversal lengths, such that $\bar{d} << \bar{L}$; the x-axis is the longitudinal axis of the body.

The uniform upstream flow has a velocity direction determined by the incidence angle θ and the azimutal angle β :

(16) $\quad \vec{U}_\infty = cos\theta\ \vec{e_x} + \sin\theta \cos\beta\ \vec{e_y} + \sin\theta \sin\beta\ \vec{e_z}$

where $\vec{e_x},\ \vec{e_y},\ \vec{e_z}$ are the unit vectors of the $x,\ y,\ z,$ axis.

The flow velocity can be decomposed into the longitudinal velocity $u\ \vec{e_x}$ and the transversal velocity $\vec{V}$:

$\vec{U} = u\ \vec{e_x} + \vec{V},\ \vec{V} = v\ \vec{e_y} + w\ \vec{e_z}$

The velocity field $\vec{V}$ in each plane $x = cst$ defines the transversal flow. Under the general conditions :

(17) $\quad |u - \cos\theta| << 1$

(18) $\quad |u - \cos\theta| \; |\partial\varphi / \partial x| << |\vec{V}| \; ((\partial\varphi / \partial y)^2 + (\partial\varphi / \partial z)^2)^{\frac{1}{2}}$

where φ represents ρ, $\vec{U}$, or e, the full Euler system can be simplified to yield the following set of equations for the transversal flow :

$$
(19) \quad \begin{cases} a)\ \cos\theta\ \partial\rho/\partial x + \mathrm{div}_T\,(\rho\vec{V}) = 0 \\ b)\ \cos\theta\ \partial\,(\rho\vec{V})\ /\ \partial x + \mathrm{div}_T\,(\rho\vec{V} \otimes \vec{V} + p\ \vec{\mathrm{I}}) = 0 \\ c)\ \cos\theta\ \partial\,(\rho E_T)\ /\ \partial x + \mathrm{div}_T\,((\rho E_T + p)\ \vec{V}) = 0 \\ d)\ E_T = e + \vec{V}^2/2, \quad e = \text{specific energy} = h/\gamma \\ e)\ p = (\gamma - 1)\ \rho\ e \end{cases}
$$

where the subscript T on the operator div_T indicates the restriction of the divergence operator to the transverse plane $(y,\ z)$.

The slip condition on the body surface, $B(x,\ \tilde{y},\ \tilde{z}) = 0$, is also approximated as :

(20) $\quad \cos\theta\ \partial B/\partial x + (v/\tau)\ \partial B/\partial\,\tilde{y} + (w/\tau)\ \partial B/\partial\,\tilde{z} = 0$

where $\tau = \bar{d}/\bar{L}$, with $\tau << 1$.

The upstream boundary conditions for the transversal flow are :

$$
(21) \quad \begin{cases} \rho = 1, \quad e = e_\infty \quad (\text{recall } e_\infty = 1/(\gamma(\gamma-1)\ M_\infty^2)) \\ v = \sin\theta\ \cos\beta, \quad w = \sin\theta\ \sin\beta. \end{cases}
$$

The system (19) is in the correct conservative form and so contains the corresponding approximate form of the shock jump relations.

By considering $(x/\cos\theta)$ as a time variable t', the above equations and boundary conditions can be identified with the exact Euler flow model for 2D unsteady flow in the (y, z) plane, past a body with moving surface given by the equation $B(t'\cos\theta, y/\tau, z/\tau) = 0$, and with uniform flow conditions at infinity given by Eqs. (21). This constitutes the hypersonic equivalence principle. In fact, this 2D unsteady Euler model, involving the time derivative with respect to t', can also be obtained as the approximate form of the exact 3D unsteady Euler model written in a reference frame moving with velocity $\bar{U}_\infty\ \cos\theta$ along the x-axis with respect to the body, t' being the dimensionless time variable in this reference frame ; so the equivalence principle corresponds in reality to a change of reference frame.

The longitudinal velocity can be obtained after the determination of the transverse flow, through the approximate Bernoulli equation :

$$
(22) \quad \begin{cases} u' = \cos\theta + u'\ , \text{ with } |u'| << 1 \\ u\prime\ \cos\theta = \gamma(e_\infty - e) - \vec{V}^2/2 + \sin^2\theta/2 \end{cases}
$$

At this point, the parameters entering into this approximate model of the transversal flow are the same as with the full Euler system, i.e. : $\gamma,\ e_\infty, \theta,\ \beta$ and τ. We will see now that we can take advantage of this model to introduce an extended similitude, but we have to distinguish two cases according to whether the incidence is small or not.

IV.2.2.2 Hypersonic small perturbation approximation

Assuming θ to be small, of the order of τ, i.e. :

(23) $\quad \theta = \tau\tilde{\theta}$, $\tilde{\theta}$ of order 1 (or small)

we can replace $cos\theta$ by 1 and $sin\theta$ by $\tau\tilde{\theta}$ in the above equations. Then we carry out the following changes of dependent and independent variables, so as to reduce the number of parameters. First, to eliminate τ from the slip condition (20), we use the following transformed coodinates and velocity components :

$$(24) \quad \begin{cases} \tilde{y} = y/\tau, \; \tilde{z} = z/\tau \\ \tilde{v} = v/\tau, \; \tilde{w} = w/\tau, \text{ i.e.} \\ \tilde{\vec{V}} = \vec{V}/\tau \end{cases}$$

In order for τ not to appear in the equations (19), we define new dependent variables :

(25) $\quad p = \tau^2\tilde{p}, \; e = \tau^2\,\tilde{e}, \; E_T = \tau^2\,\tilde{E}$

The upstream conditions (21) then write :

$$(26) \quad \begin{cases} p = 1, \; \tilde{e} = \tilde{e}_\infty = 1/(\gamma(\gamma-1)\,M_\infty^2\,\tau^2) \\ \tilde{v} = \tilde{\theta}\cos\beta\,, \; \tilde{w} = \tilde{\theta}\,sin\beta \end{cases}$$

The parameters now entering into the transversal flow model expressed in these new variables are γ, $\tilde{e}_\infty, \beta$, and $\tilde{\theta}$. The parameters e_∞ and τ are thus replaced by the parameter $\tilde{e}_\infty$. We can use, instead of $\tilde{e}_\infty$, the more classical parameter $K = M_\infty\tau$ called the hypersonic small perturbation similitude parameter.

Note that the expression (22) of u' becomes :

(27) $\quad u' = \tau^2\,(\gamma\,(\tilde{e}_\infty - \tilde{e}) - \tilde{\vec{V}}^2/2 + \tilde{\theta}^2/2)$

Similar flows can be obtained, for example, by letting $\bar{\mathcal{R}}$, $\bar{U}_\infty$, $\bar{p}_\infty, \bar{L}$ and $\bar{d}$ vary freely, while constraining the other data to vary as follows :

$\bar{C}_p \sim \bar{\mathcal{R}}, \; \theta \sim \bar{d}/\bar{L}, \; \beta\ fixed, \; \bar{e}_\infty \sim (\bar{U}_\infty\bar{d}/\bar{L})^2.$

IV.2.2.3 Slender bodies at arbitrary angle of incidence

In the case when the angle of incidence is neither small nor close to $\pi/2$, a different change of dependent variables allows to reduce the number of parameters (keeping x, $\tilde{y}$, $\tilde{z}$, as coordinates), namely :

$$(28) \quad \begin{cases} v = \sin\theta\,\tilde{\tilde{v}}, \; w = \sin\theta\,\tilde{\tilde{w}}, \text{ i.e. } \vec{V} = \sin\theta\,\tilde{\tilde{\vec{V}}} \\ p = \sin^2\theta\,\tilde{\tilde{p}}, \; e = \sin^2\theta\,\tilde{\tilde{e}}, \; E_T = \sin^2\theta\,\tilde{\tilde{E}}_T \end{cases}$$

It can be verified that the only parameter, besides γ, involved in the transformed equations and slip condition is the factor $K_1 = \tau$ cotg θ which multiplies all the x-derivative terms in the transformed equations.

The upstream conditions take the form :

$$(29) \quad \begin{cases} \rho = 1, \; \tilde{\tilde{e}} = \tilde{\tilde{e}}_\infty = 1/(\gamma\,(\gamma-1)\,M_\infty^2\,\sin^2\,\theta) \\ \tilde{\tilde{v}} = \cos\beta, \; \tilde{\tilde{w}} = \sin\beta \end{cases}$$

Thus the parameters in the transformed model are γ, β, K_1, and $\tilde{\tilde{e}}_\infty$. We can also replace $\tilde{\tilde{e}}_\infty$ by the parameter $K_2 = M_\infty \sin\theta$.

The validity of the model (19), (20) for large angle of attack in hypersonic flow and the corresponding similitude have been established by Sytchev [7] under the conditions $M_\infty \tau > 1$ and $M_\infty \sin\theta >> 1$. Barnwell [8] has relaxed the last condition to $M_\infty \sin\theta > 1$ and Hemsch [9] has shown that Sytchev's similitude also correlates well experimental data for supersonic flows with subsonic crossflow.

Notice that the parameter K_1 which multiplies the x-derivative terms in the transformed model is small, since τ is small. It can be shown that the transversal flow model corresponds to a second order approximation of the Euler model.

Similar flows can be obtained, for example, by letting $\bar{\mathcal{R}}, \bar{U}_\infty, \bar{\rho}_\infty, \bar{L}, \bar{d}$, vary freely, while imposing for the other data :

$\bar{C}_p \sim \bar{\mathcal{R}}$, $\theta \sim \text{Arctg}(\bar{d}/\bar{L})$, β fixed, $\bar{e}_\infty \sim \bar{U}_\infty^2\, \bar{d}^2\, (\bar{d}^2 + K_1^2 \bar{L}^2)$.

IV.3 Hypersonic viscous interaction

The high temperatures and low densities encountered in the viscous layers in hypersonic flow (excluding cases of strongly cooled wall) lead to thick boundary layers and to important viscous interaction effects, especially in the case of thin bodies.

The hypersonic small perturbation theory for thin or slender bodies (§IV. 2.2.2) can be associated with boundary layer theory to yield the hypersonic viscous interaction similitude which we demonstrate below (Refs. [1], [10]).

We consider plane steady laminar flows past thin bodies belonging to a family given by the equation, in cartesian coordinates :

$$(30) \qquad \bar{y} = \bar{d}\, f(\bar{x}/\bar{L})$$

where $\bar{d}$ and $\bar{L}$ are characteristic transversal and longitudinal lengths, with $\bar{d} << \bar{L}$; we define $\tau = \bar{d}\,/\bar{L}$, a small parameter ; the function $f(x)$ and its first derivative are assumed to be never large.

The viscous flow around the body is described by the boundary layer equations and the inviscid flow is described by the hypersonic small perturbation theory. However, we assume that the boundary layer thickness is not negligible compared to the body thickness, so that the inviscid slip condition must be applied to the displacement surface of equation $\bar{y} = \bar{d}\, f + \bar{\delta}^*$, where $\bar{\delta}^*$ is the boundary layer displacement thickness.

The model is given below in full, in the P.U.S., using the same notations as in § III.1.

Viscous flow

$$(31) \quad \begin{cases} a)\ \partial(\rho u)/\partial x + \partial(\rho v)/\partial z = 0 \\ b)\ \rho u\, \partial u/\partial x + \rho v\, \partial u/\partial z + dp_e\,(x)/dx = \partial(\mu\, \partial u/\partial z)/\partial z \\ c)\ \rho u\, \partial h/\partial x + \rho v\, \partial h/\partial z - u_e(x)\, dp_e(x)/dx \\ \quad = \mu(\partial u/\partial z)^2 + (1/\mathcal{R}_r)\, \partial(\mu\, \partial h/\partial z)\partial z \end{cases}$$

(32) $\rho h = (\gamma/(\gamma-1))\, p_e(x),\ T = p_e(x)/\rho$

(33) $\mu = D_\mu\, T^\omega,\ D_\mu = 1/R_e$

Here the usual intrinsic body coordinates of boundary layer theory have been assimilated with the coordinates x and $z = y - \tau f(x)$, and the tangential and normal velocity components have been assimilated with u and v (considering the upper side only). The viscosity-temperature relation is taken to be a power law, and the Reynolds number R_e is given by Eq.(21) of § III.1. The wall conditions are :

(34) $u = v = 0$ at $z = 0$

(35) $$\begin{cases} a)\ h = h_w\ (i.e.\ T = T_w), \text{or} & \\ & \text{at } z = 0 \\ b)\ \partial h/\partial z = 0 & \end{cases}$$

The outer boundary conditions are :

(36) $u \to u_e(x),\ h \to h_e(x),$ as $z \to \infty$

The displacement thickness is given by :

(37) $$\delta^* = \int_0^\infty (1 - \rho u/(\rho_e u_e))\, dz$$

Inviscid flow

The equations are derived from Eq.(19) of § IV.2.2.1 by considering θ to be small , of the order of τ ; they are written in the conservative form which includes the proper approximate shock jump relations.

(38) $$\begin{cases} a)\ \partial\rho/\partial x + \partial(\rho v)/\partial y = 0 \\ b)\ \partial(\rho v)/\partial x + \partial(\rho v^2 + p)/\partial y = 0 \\ c)\ \partial(\rho E_T)/\partial x + \partial((\rho E_T + p)v)/\partial y = 0 \end{cases}$$

(39) $E_T = e + v^2/2$

(40) $p = (\gamma-1)\,\rho e,\ h = \gamma e$

(41) $u - 1 = h_\infty - h - v^2/2$

At some distance, we recover the unperturbed flow, at an incidence θ with respect to the x-axis, hence the conditions :

(42) $$\begin{cases} \rho = 1,\ e = e_\infty = 1/(\gamma(\gamma-1)\, M_\infty^2 & \\ & y > y_1, \text{ or } x < x_0 \\ v = \theta \text{ (hence, } u = 1 - \theta^2/2) & \end{cases}$$

The slip condition on the displaced body writes :

(43) $v = \tau df(x)/dx + d\delta^*/dx$, at $y = \tau f(x) + \delta^*$

The matching between inviscid and viscous flow results from relations (36), (37) and (43) in which the subscript e refers to the inviscid values at the displacement surface :

$$(44)\quad \begin{cases} a)\ u_e(x) = u^I(x,\tau f + \delta^*),\ h_e(x) = h^I(x,\tau f + \delta^*) \\ b)\ p_e(x) = p^I(x,\tau f + \delta^*),\ \rho_e(x) = \rho^I(x,\tau f + \delta^*) \end{cases}$$

Whenever some doubt may arise on the notation, we use the superscript I for the inviscid flow solution.
At this stage, the data and the parameters of this family of flows thus modelled are the same as with the full Navier-Stokes model, as described in § III.1. However, we can try to derive an extended similitude by taking advantage of the special features of the present model.
First, for the inviscid flow, we change to the appropriate variables already introduced in § IV.2.2.2 :

$$(45)\quad \begin{cases} a)\ \tilde{y} = y/\tau,\ \tilde{v} = v/\tau,\ \tilde{h} = h/\tau^2 \\ b)\ \tilde{p} = p/\tau^2,\ \tilde{e} = e/\tau^2,\ \tilde{E}_T = E_T/\tau^2 = \tilde{e} + \tilde{v}^2/2 \end{cases}$$

so as to recover the hypersonic small perturbation similitude and to replace the two parameters τ and M_∞ (or e_∞) by the single parameter $K = M_\infty\ \tau$. From relations (43), (44), (37) we see that a necessary condition, for this to be possible, is that the displacement thickness δ^* varies as τ , at fixed K. But δ^* depends on all the parameters, in particular M_∞ and R_e, and we have to check if it is possible to have R_e vary with M_∞, at fixed K, in such a way that δ^* varies as $1/M_\infty$.

First it can be noted that the following general scaling transformation in the viscous model :

$$(46)\quad z' = z/\beta,\ v' = v/\beta,\ \beta > 0$$

amounts to replacing R_e by $\beta^2 R_e$ in Eqs. (31-b,c). The constant β remains at our disposal in view of reducing the number of parameters.

Then we have to take into account the correct order of magnitude of the boundary layer variables, i.e. to use variables which remain of constant order of magnitude as we let τ tend to zero at fixed K (hence $M_\infty \to \infty$). We know that the pressure is of order τ^2, and we write :

$$(47)\quad p_e = \tau^2\, \tilde{p}_e$$

Because of the thermal wall condition (35) (in the case (35-a), we assume that h_w remains fixed as $\tau \to 0$ and is not too small), h is of order 1 and remains so over the major part of the boundary layer thickness, whereas it is of order τ^2 in the inviscid flow. Then, in order for the condition (36) to be satisfied, h must vary rapidly from order 1 to order τ^2 in a thin transition zone at the outer edge of the boundary layer. We see from Eqs. (32) and (47) that, as a consequence, a similar but reverse situation exists for the density : ρ is of order τ^2 over the major part of the boundary layer and changes to order 1 across the same transition zone. This behaviour has been discussed by Hayes and Probstein [10] and by Cheng [11].

So, the appropriate dimensionless variables for the viscous flow are $u, h, \tilde{p}_e$ and ρ', with :

$$(48)\quad \rho' = \rho/\tau^2$$

The transformed viscous model writes as follows :

$$(49)\quad \begin{cases} a)\ \partial(\rho' u)/\partial x \ +\ \partial(\rho'\, v')/\partial z' \ =\ 0 \\ b)\ \rho' u\ \partial u/\partial x \ +\ \rho'\, v'\ \partial u/\partial z' \ +\ d\tilde{p}_e(x)/dx \\ \qquad =\ (\tau^2\ \beta^2\ R_e)^{-1}\ \partial(\mu_1\ \partial u/\partial z')\ /\partial z' \\ c)\ \rho' u\ \partial h/\partial x \ +\ \rho'\, v'\ \partial h/\partial z' \ -\ u_e\ d\tilde{p}_e(x)/dx \\ \qquad =\ (\tau^2\ \beta^2\ R_e)^{-1}\ \left[\mu_1(\partial u/\partial z')^2 \ +\ (1/\mathcal{R}_r)\ \partial(\mu_1\ \partial h/\partial z')/\partial z'\right] \end{cases}$$

$$(50)\qquad \mu_1 \ =\ T^{\omega}\ =\ [((\gamma-1)/\gamma)\ h]^{\omega}$$

$$(51)\qquad \rho' h \ =\ (\gamma/(\gamma-1))\ \tilde{p}_e(x),\ T\ =\ \tilde{p}_e(x)/\rho'$$

The boundary conditions at the wall, $z' = 0$, are practically unchanged and will not be rewritten ; the outer boundary conditions are :

$$(52)\qquad u \to u_e(x),\ h \to \tau^2\ \tilde{h}_e(x),\ \text{as } z' \to \infty$$

The displacement thickness is now given as :

$$(53)\qquad \begin{cases} \delta^* \ =\ \beta\ \delta^{*\prime}, \text{with} \\ \delta^{*\prime}\ =\ \int_0^{\infty}(1-\tau^2\ \rho' u/(\rho_e\ u_e))dz' \end{cases}$$

We impose the relation

$$(54)\qquad \tau^2\beta^2 R_e \ =\ 1$$

in order to eliminate the parameters τ and R_e from Eqs. (49), and we take

$$(55)\qquad \beta\ =\ \tau$$

since we want δ^* to vary as τ.

However, we see that the viscous model still depends on τ through the condition (52) for h, and that the inviscid model, in the transformed variables (45), also keeps a dependency on τ through the displacement effect term in the slip condition :

$$(56)\qquad \tilde{v}\ =\ df(x)/dx \ +\ d\delta^{*\prime}/dx$$

since τ appears explicity in the integrand of the expression (53) for $\delta^{*\prime}$. Therefore, we cannot rigorously extend the hypersonic small perturbation similitude to the viscous case. However, inasmuch as the remaining dependency on τ is weak, being restricted to this thin transition zone at the outer edge of the boundary layer, we obtain an approximate similitude resulting frome the relations (54) and (55). Note that the expression (53) of δ^* shows that δ^* is very close to the boundary layer thickness itself (say the value of z at which $u/u_e = 0.99$), since the integrand is equal to 1 to terms of order τ^2, over all the viscous layer except in the transition zone where ρ' becomes of order $1/\tau^2$.

Using relations (54) and (55), we obtain the following relations between the hypersonic (inviscid) similitude parameter K, the Mach number M_∞ and the Reynolds numbers R_e or R_∞ (Eq. (31) of § III.1 with $T_S = 0$) :

$$(57)\qquad K\ =\ M_\infty\,\tau\ =\ M_\infty/R_e^{1/4}\ =\ M_\infty(\gamma\ M_\infty^2)^{\omega/4}\ /R_\infty^{1/4}$$

In summary, the parameters which govern this approximate hypersonic viscous interaction similitude are the following : $\gamma, \mathcal{R}_r, \tilde{\theta}, \omega, h_w$, and K. Similar flows can be generated, for example, by letting the data $\bar{\mathcal{R}}, \bar{D}_\mu, \bar{d}, \bar{L}$, and $\bar{U}_\infty$ vary freely, while the other data vary as follows :

$\bar{C}p \sim \bar{\mathcal{R}}$, ω fixed, $\bar{D}_k \sim \bar{\mathcal{R}}$, $\bar{h}_\infty \sim \bar{U}_\infty^2 \bar{d}^2/\bar{L}^2$, $\bar{h}_w \sim \bar{U}_\infty^2$,

$\bar{\rho}_\infty \sim \bar{D}_\mu \bar{\mathcal{R}}^{-\omega} \bar{U}_\infty^{2\omega-1} \bar{L}^3 \bar{d}^{-4}$

Studies on hypersonic viscous interaction generally assume $\omega = 1$ for the viscosity law but adjust the constant so as to obtain the correct value of viscosity for temperatures in the range of wall or stagnation temperatures ; the viscosity law is written in the form :

(58) $\quad \bar{\mu} = \bar{C}_\infty \, \bar{\mu}_\infty \, \bar{T}/\bar{T}_\infty$

where $\bar{\mu}_\infty$ is the correct value of viscosity at the temperature $\bar{T}_\infty$, and where the constant $\bar{C}_\infty$ is such that $\bar{\mu}_w = \bar{C}_\infty \, \bar{\mu}_\infty \, \bar{T}_w/\bar{T}_\infty$ is the correct value at the temperature $\bar{T}_w$; then $\bar{\mu}(\bar{T}_\infty) = \bar{C}_\infty \, \bar{\mu}_\infty$ and $\bar{D}_\mu = \bar{C}_\infty \, \bar{\mu}_\infty/\bar{T}_\infty$.

The usual hypersonic viscous interaction parameter, noted $\bar{\chi}$, is defined as being proportional to K^2, and is expressed in terms of the Reynolds number R'_∞ based on the exact upstream viscosity $\bar{\mu}_\infty$:

(59) $\quad R'_\infty = \bar{\rho}_\infty \, \bar{U}_\infty \, \bar{L} \, / \bar{\mu}_\infty = \bar{C}_\infty \, R_\infty$

hence the usual form of $\bar{\chi}$:

(60) $\quad \bar{\chi} = K^2 = M_\infty^3 (\gamma \, \bar{C}_\infty / R'_\infty)^{1/2}$

V HYPERSONIC FLOWS OF REAL GASES

V.1 Equilibrium gas flows

V.1.1. General case

Similitude for flows of real gases in thermodynamic equilibrium (i.e. chemical and thermal equilibrium) is examined on the basis of the general Navier-Stokes model described in § III.1, but replacing the relations (5), (6), (9) and (10) of § III.1 by general relations of the type :

(1) $\quad \bar{p} = \bar{\rho} \, \bar{\mathcal{R}} \, \bar{T} \, Z \, (\bar{\rho}, \bar{T})$

where $\bar{\mathcal{R}}$ is some reference gas constant and Z is a dimensionless function ;

(2) $\quad \bar{h} = \bar{f}_1(\bar{\rho}, \bar{T})$

(3) $\quad \bar{\mu} = \bar{f}_2(\bar{\rho}, \bar{T})$

(4) $\quad \bar{k} = \bar{f}_3(\bar{\rho}, \bar{T})$

We do not consider the chemical composition of the gas and we neglect diffusion effects.

For practical applications, the above functions are given for a particular gas or gas mixture either in tabular form or as analytical functions involving numerical, but dimensional constants ($\bar{\mathcal{R}}$ is also one of these constants). The family $\mathcal{F}$ of flows (§ II.1) considered here is restricted to flows of this given gas or gas mixture, i.e. for given functions $Z, \bar{f}_1, \bar{f}_2, \bar{f}_3$ and given $\bar{\mathcal{R}}$.

Using the same basic reference quantities as in § III.1, the relations (1) to (4) become in the P.U.S. :

$$(5) \qquad p = \rho\, T\, Z(\rho\bar{\rho}_\infty, T\bar{U}^2_\infty/\bar{\mathcal{R}})$$

$$(6) \qquad h = \bar{f}_1(\rho\bar{\rho}_\infty, T\bar{U}^2_\infty/\bar{U}^2_\infty)$$

$$(7) \qquad \mu = \bar{f}_2(\rho\bar{\rho}_\infty, T\bar{U}^2_\infty/\bar{\mathcal{R}}) \;/\; (\bar{\rho}_\infty\bar{U}_\infty\bar{L})$$

$$(8) \qquad k = \bar{f}_3(\rho\bar{\rho}_\infty, T\bar{U}^2_\infty/\bar{\mathcal{R}}) \;/\; (\bar{\rho}_\infty\bar{U}_\infty\bar{L}\,\bar{\mathcal{R}})$$

The right-hand-sides of the above equations involve dimensionless combinations of the constants in the functions (including $\bar{\mathcal{R}}$) with the data $\bar{\rho}_\infty, \bar{U}_\infty, \bar{L}$; these combinations must remain fixed for similitude, and therefore, with functions $Z, \bar{f}_1, \bar{f}_2, \bar{f}_3$ having no special property, the data $\bar{\rho}_\infty, \bar{U}_\infty$, and $\bar{L}$ must remain fixed for similar flows of a fixed gas.

The remaining similitude parameters are just h_∞, T_w and geometrical parameters. Since $\bar{U}_\infty$ is fixed, similitude imposes also that $\bar{h}_\infty$ and $\bar{T}_w$ be fixed. We see that all the data must be fixed, so that there cannot be two different similar flows. Similitude is not possible for viscous gas flows of a given arbitrary gas at thermodynamic equilibrium. In the hypersonic limit, where $h_\infty \to 0$, the flow becomes independent of the parameter h_∞, and we can have approximate similitude for flows with different (but small enough)values of $\bar{h}_\infty$ (i.e. small values of $\bar{T}_\infty$), at fixed $\bar{\rho}_\infty, \bar{U}_\infty, \bar{L}$ and $\bar{T}_w$.

For inviscid flows (Euler model) of equilibrium gas, the relations (3), (4) (7), (8) disappear, and we see that $\bar{L}$ becomes free to vary. Flows differing only by the size of the body are similar (assuming the body size not too small, in order for the equilibrium model to be valid). The hypersonic small perturbation similitude (§ IV.2.2.2, see also Ref. [12]), and Sytchev's similitude (§ IV.2.2.3) can be applied with an arbitrary equilibrium gas, but with the important restriction that $\bar{U}_\infty$ and $\bar{\rho}_\infty$ should be fixed.

V.1.2. Thermally perfect gas

For a thermally perfect gas, we have :

$$(9) \qquad Z = 1$$

$$(10) \qquad h = \bar{f}_1\,(T\bar{U}^2_\infty \,/\, \bar{\mathcal{R}}) \;/\; \bar{U}^2_\infty$$

and we assume $\bar{\mu}$ and $\bar{k}$ to be functions of $\bar{T}$ only :

$$(11) \qquad \mu = \bar{f}_2(T\bar{U}^2_\infty \,/\, \bar{\mathcal{R}}) \;/\; (\bar{\rho}_\infty\bar{U}_\infty\bar{L})$$

$$(12) \qquad k = \bar{f}_3(T\bar{U}^2_\infty \,/\, \bar{\mathcal{R}}) \;/\; (\bar{\rho}_\infty\bar{U}_\infty\bar{L}_\infty\bar{\mathcal{R}})$$

We see from relations (9) to (12) that similitude of viscous flows of a given thermally perfect gas requires that $\bar{U}_\infty$ and $(\bar{\rho}_\infty\bar{L})$ be kept fixed ; hence $\bar{h}_\infty$ and $\bar{T}_w$ must also remain fixed. Compared to the general case above, we have one degree of freedom with free variations of $\bar{\rho}_\infty$, and with $\bar{L}$ varying as $1/\bar{\rho}_\infty$.

We thus get the same similitude as in the case of a fixed calorically perfect gas (§ III.1). In the hypersonic limit, $\bar{h}_\infty$ becomes also free.

For inviscid flows, $\bar{\rho}_\infty$ and $\bar{L}$ can vary independently.

V.1.3. Diatomic dissociating gas

An interesting and relatively simple real gas behaviour is that of the pure diatomic dissociating gas ; it is a particular case, but quite representative of phenomena in practical problems.

Let A_2 be the diatomic molecule, $\bar{\mathcal{R}}$ the undissociated gas constant, and α the mass fraction of monatomic gas A in the mixture : $\alpha = \rho_A/\rho$. The function Z of Eq. (1) is given by the law of mass action through the relations :

(13) $\quad Z = 1 + \alpha$

(14) $\quad \alpha^2 \;/\; (1-\alpha) = (\bar{\rho}_D(\bar{T}) \;/\; \bar{\rho}) \exp(-\bar{T}_D/\bar{T})$

where $\bar{T}_D$ is a characteristic temperature for dissociation, and where the function $\bar{\rho}_D$ is determined from the partition functions associated with the translational, rotational, vibrational and electronic degrees of freedom. Lighthill [13, 14], has discussed this gas model in detail and has proposed an approximate form, in which $\bar{\rho}_D$ is a constant, called the ideal dissociating gas. With the assumption that the electronic partition functions are constant, we get the following form of the function :

(15) $\quad \bar{\rho}_D(\bar{T}) = \bar{\rho}_{D0}\, (\bar{T}/\bar{T}_V)^{1/2}\, (1 - \exp(-\bar{T}_V/\bar{T}))$

where the constant $\bar{\rho}_{D0}$ is a characteristic density, and $\bar{T}_V$ a characteristic temperature for vibration of A_2. The enthalpy is also given by :

(16) $\quad \bar{h} = \bar{\mathcal{R}}\,\bar{T}\, \left[4 + \alpha - (1-\alpha)\,(\bar{T}/\bar{\rho}_D)\, d\bar{\rho}_D(\bar{T})/d\bar{T}\right] + \alpha\bar{\mathcal{R}}\,\bar{T}_D$

Lighthill [13] has shown that for oxygen and nitrogen, $\bar{\rho}_D(\bar{T})$ does not vary too much, so that the assumption that $\bar{\rho}_D$ is a constant, if not quantitatively accurate, is quite valuable for simplified studies. For the present purpose limited to similitude considerations, we keep the general expressions (15) and (16). The physical data involved in this gas model are $\bar{\mathcal{R}}, \bar{\rho}_{D0}, \bar{T}_D$, and $\bar{T}_V$. With Lighthill's model, $\bar{T}_V$ disappears.

In the dimensionless form of the P.U.S., the relations (13) to (16) remain formally the same, except that all variables or constants are unbarred, and that we have $\mathcal{R} = 1$. So we identify three parameters in these relations, namely :

(17) $\quad \rho_{D0} = \bar{\rho}_{D0}/\bar{\rho}_\infty,\; T_D = \bar{\mathcal{R}}\,\bar{T}_D/\bar{U}_\infty^2,\; T_V = \bar{\mathcal{R}}\,\bar{T}_V/\bar{U}_\infty^2$

The similitude conditions for flows of a given gas are the same as in the general case (§ V.1.1.). For inviscid flows, $\bar{L}$ is free, and if we allow variations of the gas data, then similar flows of two different gases, referred to with superscripts (1) and (2), will be obtained subject to the following constraints :

(18) $\quad \bar{T}_D^{(2)}/\bar{T}_V^{(2)} = \bar{T}_D^{(1)}/\bar{T}_V^{(1)}$

(19) $\quad (\bar{U}_\infty^{(2)}/\bar{U}_\infty^{(1)})^2 = \bar{h}_\infty^{(2)}/\bar{h}_\infty^{(1)} = \bar{\mathcal{R}}^{(2)}\,\bar{T}_D^{(2)}/\bar{\mathcal{R}}^{(1)}\,\bar{T}_D^{(1)}$

(20) $\quad \bar{\rho}_\infty^{(2)}/\bar{\rho}_\infty^{(1)} = \bar{\rho}_{D0}^{(2)}/\bar{\rho}_{D0}^{(1)}$

For example, comparing data for O_2 and N_2 (Ref. [13]), we find :

$$(\bar{T}_D/\bar{T}_V)_{O_2} = 26.5, \quad (\bar{T}_D/\bar{T}_V)_{N_2} = 33.8$$

Approximately similar flows of dissociating O_2 and of dissociating N_2 (at equilibrium) will be obtained if we have the following relations :

$$\bar{\rho}_\infty(O_2)/\bar{\rho}_\infty(N_2) = 1.24, \; \bar{U}_\infty(O_2)/\bar{U}_\infty(N_2) = 0.68,$$
$$\bar{h}_\infty(O_2)/\bar{h}_\infty(N_2) = 0.46.$$

With Lighthill's gas model, the constraint (18) disappears.

V.2 Non-equilibrium reacting gas flows

With the complex models presently used for the computation of real non-equilibrium gas flows, including vibrational and chemical non-equilibrium , similitude is not possible. The only approximate similitude is the one obtained in the hypersonic limit, i.e. when $\bar{h}_\infty$ tends to zero at fixed $\bar{\rho}_\infty, \bar{U}_\infty, \bar{T}_w$ and $\bar{L}$. It is a Mach number independence principle in which, as stressed by Hayes and Probstein [2], similar flows differ only in the freestream by the temperature and pressure values, and are practically identical in the perturbed flow region.

However approximate similitude may be obtained with simple models. A classical and important example is provided by a dissociating diatomic gas, when recombination can be neglected compared to dissociation. The ideal dissociating gas model of Lighthill has been extended to the case of chemical non-equilibrium by Freeman [15], and this extended model has been used in the past for various studies (e.g. Refs. [15] to [23]). We now briefly discuss this model, assuming inviscid flows.

The thermodynamic relations (1), (13), (15) and (16) remain valid, but now the mass fraction α must be determined from a rate equation. Freeman proposes the following equation :

$$(21) \qquad \vec{\bar{U}}.\operatorname{grad}\alpha = \bar{C}_1\,\bar{\rho}\,\bar{T}^{-s}\,\left[(1-\alpha)\exp(-\bar{T}_D/\bar{T}) - \alpha^2\bar{\rho}/\bar{\rho}_D\right]$$

where $\bar{C}_1, s$, and $\bar{\rho}_D$ are constants. Here we can let $\bar{\rho}_D$ vary according to the formula (15), which implies that the vibration of molecules is at equilibrium with the translational temperature.

In dimensionless form (in the P.U.S.), Eq. (21) becomes :

$$(22) \qquad \vec{U}.\operatorname{grad}\alpha = (\bar{C}_1\,\bar{\rho}_\infty\,\bar{L}/\bar{U}_\infty)\,(\bar{U}^2_\infty/\bar{\mathcal{R}}^{-s})\,\rho T^{-s}\,[(1-\alpha)\exp(-T_D/T) - \alpha^2\rho/\rho_D]$$

Besides the parameters given by Eq. (17), we now have the new parameter :

$$(23) \qquad \chi = (\bar{C}_1\,\bar{\rho}_\infty\,\bar{L}/\bar{U}_\infty)\,(\bar{U}^2_\infty/\bar{\mathcal{R}})^{-s}$$

For a fixed gas, similitude of inviscid flows imposes that $\bar{\rho}_\infty$ and $\bar{U}_\infty$ be fixed (from the parameters (17)), and therefore we see from the expression (23) of the parameter χ that $\bar{L}$ must also be fixed. Equation (21) introduces a characteristic time, $\bar{t}_R$, hence a characteristic length $\bar{L}_R = \bar{U}_\infty\bar{t}_R$, and the parameter χ is equal to the ratio $\bar{L}/\bar{L}_R$, the inverse of the Damkohler number. When $\chi \to \infty$, the material derivative of α remains bounded only if the last parenthesis in the R.H.S. of (22) tends to zero, which yields the equilibrium relation (14) ; on the contrary, if $\chi \to 0$, then α remains constant on each streamline (frozen flow).

There may be situations, like just behind a strong shock, where dissociation dominates over recombination i.e., referring to Eq. (22), where :

$$(24) \qquad \alpha^2 \rho / \bar{\rho}_D << (1-\alpha)\exp(-T_D/T)$$

In such cases, we approximate Eq. (22) by :

$$(25) \qquad \vec{U}.\mathrm{grad}\alpha \;=\; \chi\, \rho\, T^{-s}(1-\alpha)\exp(-T_D/T)$$

Then the function $\rho_D(T)$ and the parameter ρ_{D0} do not appear in the model, and similitude does not require $\bar{\rho}_\infty$ to be fixed. The constancy of the parameter χ is then satisfied with ($\bar{\rho}_\infty \bar{L}$) fixed (recall that $\bar{U}_\infty$ is fixed), and we can have different similar inviscid flows of the same gas, with same $\bar{U}_\infty$ and with the condition that $\bar{\rho}_\infty$ varies as $1/\bar{L}$. This similitude is referred to as the binary scaling, since it comes from the fact that dissociation involves binary molecular collisions, giving a linear dependency on ρ in Eq. (22), whereas recombination involves tertiary collisions leading to a quadratic dependency on ρ.

CONCLUSION

Hypersonic aerodynamics has provided a strong incentive, in the past, to the study of approximate flow models and of the related similitudes. Some of the most classical hypersonic flow models have been reviewed here, especially in view of exhibiting the similitude which these models lead to. An important subject which has not been covered here is the study of bluntness effects in slender body flows, e.g. Ref. [24].

Hypersonics is again in the forefront of research in aerodynamics, and a large effort is devoted to the development of numerical methods, in particular for non-equilibrium real gas flows. Despite the great advances in the numerical solution of the full Euler or Navier-Stokes equations, the existing approximate models for hypersonic flows should remain of great interest, not only from a fundamental point of view because of the insight which they provide into the basic phenomena, but also from a practical point of view because of the reduced computing cost and of the similitude laws which they provide.

A difficulty usually encountered when using an approximate model concerns the determination of its domain of validity for a maximum accepted error with respect to a more complete model taken as a reference. Computational aerodynamics provides the means to solve this difficulty by comparisons of accurate numerical solutions of the approximate and of the reference models. For example, the hypersonic small perturbation model can be checked against the full Euler model.

Also, it is possible to determine the accuracy and the practical domain of validity of an approximate similitude for a given model, such as the Mach number independence principle or the hypersonic viscous similitude, by testing this similitude on accurate numerical solutions of the model.

These capabilities afforded by the progress in computational aerodynamics should provide a renewed interest in approximate models and in the study of approximate similitudes.

Acknowledgments

The author expresses his thanks to J.P. Guiraud for helpful comments.

REFERENCES

[1] HAYES, W.D., and PROBSTEIN, R.F. - Hypersonic Flow Theory, Academic Press, 1959.

[2] HAYES, W.D., and PROBSTEIN, R.F. - Hypersonic Flow Theory, 2nd Ed., Vol I, Inviscid Flows, Academic Press, 1966.

[3] SEDOV, L.I. - Similarity and Dimensional Methods in Mechanics, Infosearch Ltd., London, 1959.

[4] MARTINOT-LAGARDE, A. - Analyse Dimensionnelle. Applications à la Mécanique des Fluides, ONERA Rapport Technique $N°$ 41, 1948.

[5] GUIRAUD, J.P. - Newtonian Flow Over a Surface - Theory and Applications,in Vol. XI of the Colston Papers", Butterworths Scientific Publ., London, 1959.

[6] GUIRAUD, J.P. - Ecoulements Hypersoniques Tridimensionnels dans une Couche de Choc d'Epaisseur Evanescente, La Rech. Aéron. $N°$71, Juil.-Aout 1959.

[7] SYCHEV, V.V. - Three-Dimensional Hypersonic Gas Flow Past Slender Bodies at High Angles of Attack, J. of Appl. Math. and Mech. (PMM), Vol 24, $N°$2, 1960.

[8] BARNWELL, R.W. - Extension of Hypersonic High-Incidence, Slender-Body Similary, AIAA J., Vol. 25, $N°$11, Nov.1987.

[9] HEMSCH, M.J. - Engineering Analysis of Slender Body Aerodynamics Using Sychev Similarity Parameters, J. of Aircraft, Vol.25, $N°$7, July 1988.

[10] HAYES, W.D., and PROBSTEIN, R.F. - Viscous Hypersonic Similitude, J. Aerosp. Sc., Vol.26, $N°$12, Dec. 1959.

[11] CHENG, H.K. - Hypersonic Gas Dynamics of Slender Bodies, in "Modern Developments in Gas Dynamics", Plenum Press, 1969.

[12] CHENG, H.K. - Similitude of Hypersonic Real-Gas Flows Over Slender Bodies With Blunted Noses, J. Aerosp. Sc., Vol.26, $N°$9, Sept. 1959.

[13] LIGHTHILL, M.J. - Dynamics of a dissociating gas. Part 1 Equilibrium flow, J. Fluid Mech., Vol.2, Part 1, Jan. 1957.

[14] LIGHTHILL, M.J. - Dynamics of a dissociating gas. Part 2 Quasi-equilibrium transfer theory, J. Fluid Mech., Vol.8, Part 2, June 1960.

[15] FREEMAN, N.C. - Non-equilibrium flow of an ideal dissociating gas, J. Fluid Mech., Vol.4, Part 4, Aug. 1958.

[16] BRAY, K.N.C. - Atomic recombination in a hypersonic wind-tunnel nozzle, J. Fluid Mech., Vol.6, Part 1, July 1959.

[17] BLOOM, M.H., and STEIGER, M.H. - Inviscid Flow with Non-Equilibrium Molecular Dissociation for Pressure Distributions Encountered in Hypersonic Flight, J.Aerosp. Sc., Vol.27, $N°$11, Nov. 1960.

[18] DORRANCE, W.H. - On the Approach to Chemical and Vibrational Equilibrium Behind a Strong Normal Shock Wave, J. Aerosp. Sc., Vol.28, $N°$1, Jan. 1961.

[19] LI, T.Y. - Recent Advances in Nonequilibrium Dissociating Gasdynamics, ARS J., Vol.31, $N°$ 2, Feb. 1961.

[20] HALL, J.G., ESCHENROEDER, A.Q., and MARRONE, P.V. - Blunt-Nose Inviscid Airflows With Coupled Nonequilibrium Processes, J. Aerosp. Sc., Vol.29, $N°$9, Sept. 1962.

[21] GIBSON, W.E., and MARRONE, P.V. - A Similitude for Non-Equilibrium Phenomena in Hypersonic Flight, in "The High Temperature Aspects of Hypersonic Flow", AGARDograph 68, NELSON W.C., Ed., Pergamon Press, 1964.

[22] GIBSON, W.E. - Dissociating Scaling for Nonequilibrium Blunt Nose Flow, ARS J., Vol.32, $N°$2, Feb. 1962.

[23] HORNUNG, H.G. - Non-equilibrium ideal-gas dissociation after a curved shock wave, J. Fluid Mech., Vol.74, part 1, 1976.

[24] GUIRAUD, J.P., VALLEE, D., and ZOLVER, R. - Bluntness Effects in Hypersonic Small Disturbance Theory, in Basic Develop. in Fluid Dyn. Vol.1, Academic Press, 1965.

GAS-KINETICAL and NAVIER-STOKES SIMULATIONS of REENTRY FLOWS

H. OERTEL jr.

Institute for Fluid Mechanics
Technical University of Braunschweig

1. INTRODUCTION

In addition to the classic wind tunnel, numerical methods are gaining more and more importance as a new tool for the development of aerospace technologies. However, the present gaskinetical and Navier-Stokes methods, which have been developed worldwide, are unsuitable for practical aerospace applications without the intermediate step of validation. This is due partly to the inadequate mathematical verification of the numerical algorithms and boundary conditions and especially due to the lack of physical modelling of nonequilibrium chemistry in the high atmosphere as well as three-dimensional boundary layer transition and turbulence modelling of chemically reacting flows.

Since aerothermodynamics at real reentry conditions can basically not be fully simulated in the wind tunnel, quantitative capsular reentry experiments must be performed in order to validate the numerical methods. From the dimension analysis in fig. 1, it is clear that a partial experimental simulation of the M_∞ - Re_∞ dependency, or rather the simulation of isolated chemical reactions in supersonic or high enthalpy tunnels is possible. However, the reentry problem in its entirety, i.e. the three-dimensional aerothermodynamics around the hot structure of the heat-shield including the chemistry of the high atmosphere in thermodynamic non-equilibrium and their interaction with the technical surface of the heat-shield, cannot be experimentally simulated.

Due to the complexity of the aerodynamic phenomena mentioned, it is of utmost importance to select a well-defined boundary-value problem for

Dimension Analysis

$\dot{q} = \dot{q}\,(Kn,\ M_\infty,\ Re_\infty,\ Pr_\infty,\ \kappa,\ T_W/T_\infty,\ l/L\ \ldots)$

Mach-Reynolds Simulation

$\dot{q} = \dot{q}\,(M_\infty,\ Re_\infty,\ T_W/T_\infty)$

Dissociation Simulation

$\dot{q} = \dot{q}\,(Re_\infty,\ T_W/T_\infty,\ l/L)$

➧ Reentry Free Flight Experiment

Fig. 1: Dimension analysis

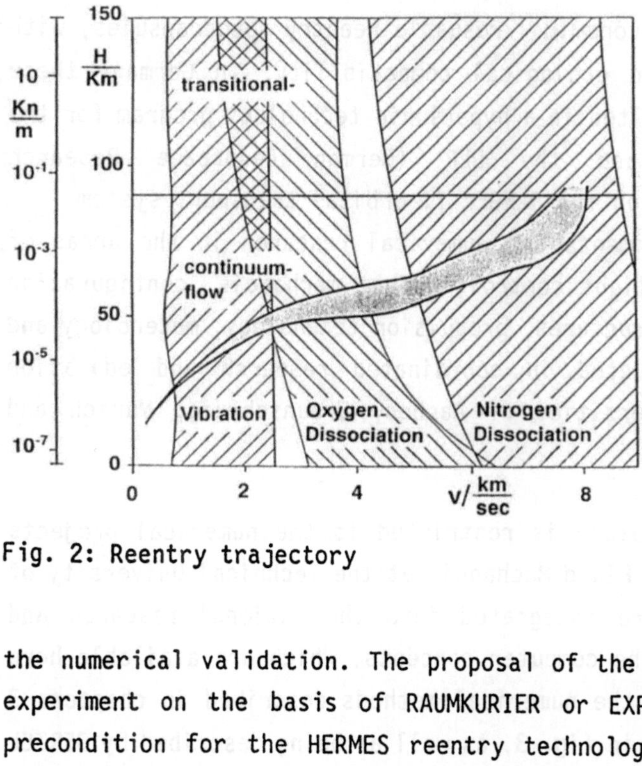

Fig. 2: Reentry trajectory

the numerical validation. The proposal of the quantitative reentry experiment on the basis of RAUMKURIER or EXPRESS, should be the precondition for the HERMES reentry technology developments. This article will introduce, after a short description of the national hypersonic programs [1], the numerical preparations of the free-flight validation experiment, which corresponding to fig. 2, makes along the reentry trajectory the unsteady simulation in the regime of the free molecular flow over the transitional regime to continuum mechanics necessary. This leads to the introductory chapters on gas-kinetical and Navier-Stokes simulations.

This paper will also partly summarize the research on the aerothermodynamic reentry efforts in Germany as seen from the perspective of the Aerospace Center at the University of Braunschweig and will conclude with a proposal for a quantitative reentry experiment to validate hypersonic wind tunnels and numerical design tools for a reentry capsula design.

2. NATIONAL HYPERSONIC PROGRAMS AND COMPUTER CONCEPTS

In connection with the scenario of newly projected spacestations comes the further development of reentry technology with special reference to the need for more economical resuable reentry spacecapsules, with optimized payloads and ecological compatibility. In Germany these requirements have resulted in a hypersonic technology program for the aerospace industry and the DLR (German Aerospace Research Establishment), based on the SÄNGER II orbital transport system.
The fundamental experiments and numerical research in the areas of aerothermodynamics, flight control, flight mechancis, configuration and system studies, structures, propulsion technology, meterology and environment are conducted in coordinated research and education programs at the universities of Aachen, Braunschweig, Munich and Stuttgart.

The remainder of the paper is restricted to the numerical projects from the Institute for Fluid Mechancis at the Technical University of Braunschweig, which are integrated into the national research and technology programs. The computer concepts, which are available here for the validation of the numerical methods described in chapters 3 and 4, are illustrated in fig. 3, as well as being described by OERTEL [2]: workstations for pre- and postprocessing with the interactive use of a vector computer for grid generation and data evaluation, a high performance data network leads to a super computer permitting parameter studies and configuration optimizations in vector computers or parallel computers. At the Technical University of Braunschweig, the sketched computer concept for hypersonic technology developments is currently being realized.

3. GAS-KINETICAL SIMULATION

3.1. Monte Carlo Simualtion (DSMC)

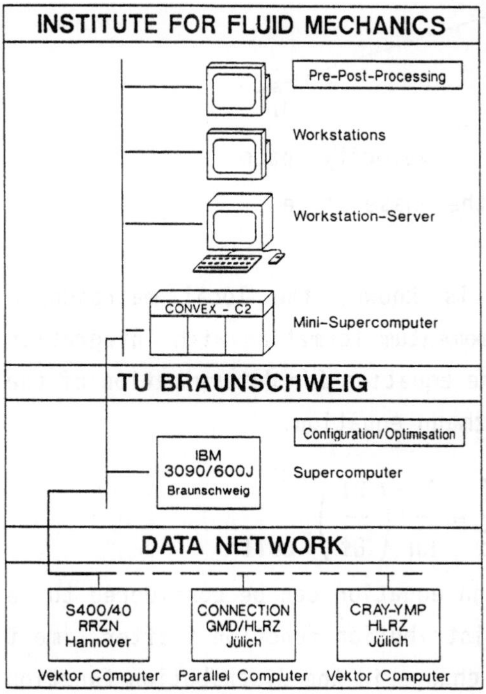

Fig. 3: Computer Concept

To begin with, as in the reentry trajectory of fig. 2, in the regime of free molecular flow and in the transitional regime to continuum mechancis the gas-kinetical simulation of the Boltzmann equation have to be developed. Moreover, we will follow the work of BIRD [3,4] MOSS and CUBA [5], and WETZEL and OERTEL [6,7].

Here, f $(\vec{r},\vec{c},t)$ represents the distribution function which is the basis for the microscopic description of flows. It describes the statistical temporal and spatial distribution of the particles and their velocities in the physical space shown in fig. 4. Together, these form the socalled phase space, which in general has six dimensions. $f(\vec{r},\vec{c},t)d\vec{c}\ d\vec{r}$ represents the number of particles positioned around $\vec{r}$ within the volume element dx dy dz and whose velocities lies in the range du dv dw around $\vec{c}$. In a state of thermodynamic equilibrium, the distribution of the velocities in the velocity space at a given temperature corresponds to the Maxwellian distribution.

Integration of the distribution function over the velocity space $\int f(\vec{r},\vec{c},t)d\vec{c}$ results in the particle density in the volume element of the physical space. Further integration over the entire physical space $\iint f(\vec{r},\vec{c},t)d\vec{c}\ d\vec{r}$ then gives the total number of particles. If the

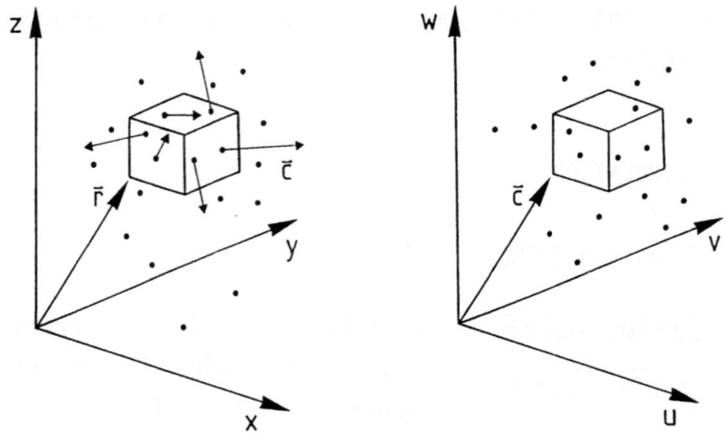

Fig. 4: Particle distribution in the phase space

molecular velocity distribution is known, the local macroscopic parameters can be calculated by momentum formation with integration over the entire velocity space. The equation for determination of the distribution function is the Boltzmann equation.

$$\frac{\partial f}{\partial t} + \bar{c}\,\frac{\partial f}{\partial \bar{r}} + \bar{F}\,\frac{\partial f}{\partial \bar{c}} = \frac{1}{Kn}\left(\frac{\partial f}{\partial t}\right)_{coll}$$

The left-hand side of the Boltzmann equation can be considered to be a substantial derivative of the distribution function f after time t in the phase space. This temporal change in the distribution function f is the consequence of intermolecular collisions, the effect of which is described by the collsion term on the right-hand side. The term $\bar{c}\,(\partial f/\partial \bar{r})$ describes the convection of the particles into the physical space; in a similar way, $\bar{F}\,(\partial f/\partial \bar{c})$ accounts for the convection of the particles into the velocity space as a result of external forces. If binary collisions are assumed, the collision term can be replaced by an integral expression, this leading to a nonlinear intergo-differential equation.

For a general, three-dimensional flow the distribution function f depends on three space coordinates and three velocity components, meaning that f must be solved for six dimensions with given initial and boundary conditions. At the moment, this is only possible for few special cases. Consequently, direct gaskinetical simulation methods have to be relied on.

In these methods, the distribution function is represented directly by the gas particles in the phase space; space coordinates and velocity components being assigned to each particle. The real gas flow with 10^{19} particles per cubic centimetre is simulated by several thousand model particles for this purpose. This is possible, since the Boltzmann equation can be normalized in such a way that the solution is independent of the number of particles. It must merely be ensured that th product $n \cdot \sigma$ is kept constant, n being the particle density and σ the cross-section of the gas particles. Modelling of the particle collision and application of the fundamental gas-kinetical equations to the model particles allows simulation of the temporal and spatial change in the distribution function f and, consequently, the corresponding change of state in the flow.

Let us consider the collision of two particles in fig. 5. The collision of particle 1 and particle 2 is observed in the center of mass frame of reference. The collision plane is generated with two particle mid-points and the relative velocity vector $\vec{c}_r$ of the two colliding particles before the collision.

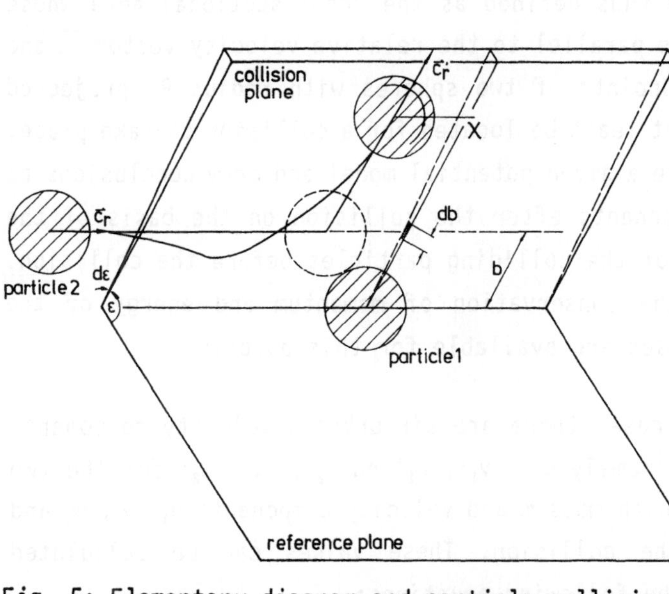

Fig. 5: Elementary diagram and particle collision

The smallest distance between particle 1 and the undisturbed path of particle 2 establishes the first collision parameter b.

With a given sphere radius and a known collision parameter b, the potential of the interaction force for rigid spheres makes it possible to calculate easily the change in direction χ between the relative velocity vectors $\vec{c}_r$, before the collision, and $\vec{c}_r'$ after the collision, in the collision plane.

A reference plane must then be introduced for unambiguous localization of the collision in space. The line intersecting this reference plane and the collision plane includes the mid-point of particle 1 and follows parallel to the relative velocity vector $\vec{c}_r$. The reference plane is defined by a randomly selected azimuth angle ϵ between collision plane and reference plane, this angle representing the second impact parameter in binary collision.
Integration over the impact parameter b and ϵ yields the cross-section σ for the rigid sphere potential model

$$\sigma = \int_0^{2\pi} \int_0^{2R} b \, db \, d\epsilon = 4 \cdot \pi R^2$$

The cross-section is thus defined as the cross-sectional aera whose surface normal $\vec{n}$ runs parallel to the relative velocity vector $\vec{c}_r$ and within which the midpoints of two spheres with radius R, projected parallel to $\vec{n}$, must at least be located for a collision to take place. The task is now to use a given potential model and draw conclusions as to the velocity components after the collision on the basis of the velocity components of the colliding particles before the collision. The equations for the conservation of momentum and energy of the kinetic theory of gases are available for this purpose.

In the most general case, there are six unknown velocity components after the collision, namely u_1', v_1', $w_1'$and u_2', v_2', w_2' for the two colliding particles with mass m and velocity components u_1, v_1, w_1 and u_2, v_2, w_2 before the collision. These values can be calculated unambiguously with the following equations:

Conservation of linear momentum per mass:

$$u_1 + u_2 = u_1' + u_2'$$
$$v_1 + v_2 = v_1' + v_2'$$
$$w_1 + w_2 = w_1' + w_2'$$

Conservation of energy per mass:

$$u_1^2 + v_1^2 + w_1^2 + u_2^2 + v_2^2 + w_2^2 =$$
$$u_1'^2 + v_1'^2 + w_1'^2 + u_2'^2 + v_2'^2 + w_2'^2$$

Conservation of angular momentum per mass:

$$\bar{r}_1 \times \bar{c}_1 + \bar{r}_2 \times \bar{c}_2 = \bar{r}_1' \times \bar{c}_1' + \bar{r}_2' \times \bar{c}_2'$$

Interaction potential:

$$m \cdot \ddot{\vec{r}}_1 = -m \cdot \ddot{\vec{r}}_2 = -\frac{d\phi}{ds} .$$

The radius vectors of the particles before and after the collision are presented by the parameters $\vec{r}_1$, $\vec{r}_2$, and $\vec{r}_1'$, $\vec{r}_2'$, respectively, in accordance with fig. 5, while s represents the distance between the colliding particles. Since, in the case of an elastic collision of rigid spheres, the impact parameters ϵ and b are obtained immediately from the geometrical consideration, thus meaning that the conservation of angular momentum is automatically fulfilled, there remain equations for unambigous determination of the velocity components after the collision.

The simulation method which we use is the Direct Simulation Monte Carlo Method (DSMC) which treats the particle collisions statistically. The simulation domain is discretized in cells. The cell size must be selected in such a way that only a slight change in the state variable is to be expected within a cell. Positioning und velocity distribution of the particles are the same as for the molecular dynamics method. The boundary condition is obtained from the geometry of the simulated flow and the given interaction between the gas particles and the walls present. The simulation process of two steps, which can be applied repeatedly:

In the first step the particles are moved with their individual velocities over the discrete time step Δt_m, this motion being decoupled from the interaction of the particles. Collisions with the walls are simulated in a way similar to the molecular dynamics method. In the second stage, representative collisions with randomly selected particles are subsequently calculated in each cell.

For the rigid sphere model, the calculation of the collision time Δt_{coll} of an individual collision process is based on the relationship

$$\Delta t_{coll} = \frac{2}{N \cdot n \cdot \pi d^2 \cdot c_r}$$

which produces correct collision rates. In this equation, N is the number of particles per cell, πd^2 the cross-section of the particles, n the particle density and c_r the relative velocity between the particles involved in the collision.

The representative collisions are carried out by randomly selecting of two particles in every cell to form a possible collision pair. Whether or not this pair collides depends on the probability of collision, this being revers proportional to the collision time Δt_{coll}. Once a suitable pair of particles has been found, the equations of motion together with two impact parameters are used to calculate the new velocity components. Collisions then take place in the same cell until the total of the time intervals Δt_{coll} reaches the selected time interval Δt_m. The time interval selected should be small in comparison with the mean collision time of the particles.

Since the particles do not really collide in this method, the impact parameters b and ϵ cannot be determined from geometrical considerations, meaining that the angular momentum equation is not automatically fulfilled.

The Direct Simulation Monte Carlo method makes use of the fact that in a statistical sense, for the hard sphere particle potential, there is no preferred direction for the post-collision relative velocity of the two particles. Consequently, an azimuth angle ϵ and a polar angle χ are determined stochastically. Here lies the problem in relation to the conservation of angular monemtum, which needs not be explicitly taken into account.

3.2 Diatomic Simulation Models

The gas viscosity in the DSMC calculations is simulated by using the empirical variable hard sphere (VHS) collision model. In this model the temperature dependency of the viscosity is considered by a variable cross section that is inversely proportional to the relative collision energy between the colliding particles.

The internal energies of molecules have been considered with the phenomenological model introduced by BORGNAKKE and LARSEN [8]. The essential feature of this model is that a part of the collisions is treated as completely inelastic. The probability for inelastic collisions, where energy transfer between translational and rotational energies may occur, is directly related to the rotational relaxation time. For the temperature range on our calculations, the rotational energy is always fully excited. At higher temperature the vibrational energy levels become excited. To predict the degree of excitation at a particular temperature which is not originally defined, the harmonic oscillator model has been used. The knowledge of the vibrational degrees of freedom enables the consideration of the translational-rotational and vibrational energy exchange in a modified version of the Borknakke-Larsen model.

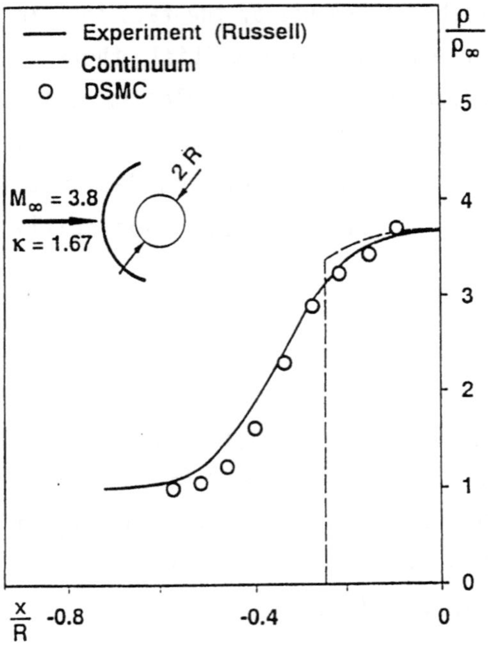

Fig. 6:
Experimental and theoretical density profiles on the stagnationline of a sphere

The proportion of such inelastic collisions is adjusted to the vibrational relaxation time.
For the verification of the aerodynamic part of the DSMC method the supersonic flow past a sphere without chemical reactions was simulated. Fig. 6 shows a comparison of the calculated results with electron beam measurements by RUSSEL [9] for the density distribution along the stagnation streamline. The flowfield conditions are $M_\infty = 3.8$, $Kn_\infty = \lambda/R = 0.06$. At the body surface the adiabatic wall temperature was fixed. The predicted 3D DSMC results are in a good agreement with the experimentally estimated density distribution ahead of the sphere.

Fig. 7 shows in addition the Mach-isolines in the calculated 3D-flowfield with a maximum of translation of 23.000 K. The sonic line that separates the subsonic and supersonic region looks different in comparision to the continuum case. The calculated subsonic region extends around the front part of the sphere to the outflow boundary, emphasizing the low density effect on the flow structure.

In the next step, the chemistry of the higher atmosphere with 19 reactions following BORTNER [10], were implimented into the Monte-Carlo method. In order to compute the probability of an inelastic collision, with rotation and vibration excitation and dissociation of oxygen and nitrogen molecules, the empirical Borgnakke-Larsen model [11] was applied. This model brings the probability of inelastic collisions into relationship with the classical rotational and vibrational degrees of freedom.

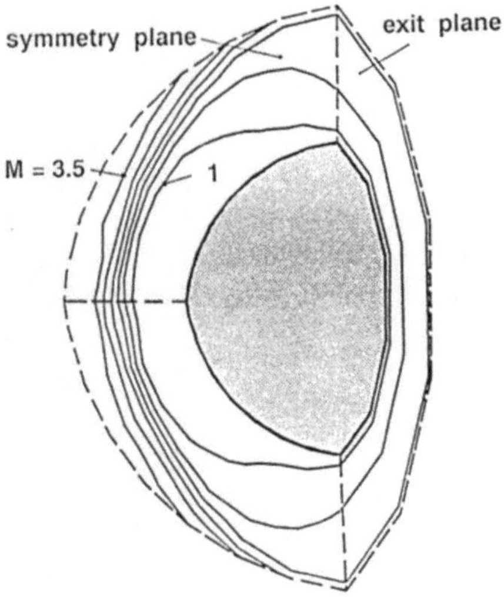

Fig. 7: Numerical iso Mach-lines in the symmetry line

This further simulation was carried out at a Knudsen number of 0.08, which corresponds to a reentry altitude of 92 km. The viscosity was simulated with the VHS-model, the rotation relaxation length was fixed at a five mean free path $\bar{\lambda}$ and the vibration relaxation length at $100\bar{\lambda}$. The isothermal wall of the sphere has a constant temperature of 1000K, as well an ideal thermal accomodation and diffuse reflection at the surface have been assumed, which as we know, is not the technical reality. Isothermal areas in front of sphere are shown in in fig. 8. The upper half of the figure is the result of a perfect gas calculation; the lower half illustrates the influence of real gas effects, with a decrease of the shock stand-off distance and the maximum temperature in front of the sphere.

In addition, the temperature components along the stagnation streamline of fig. 9, show the temperature increase in the shock front and after having reached the maximum temperature, the largest temperature decrease to the isothermal boundary conditions of 1000K are attained.

The maximum temperature in front of the body is in a real reentry case, approximately half as large as under perfect gas conditions. This shows, how important the modulation of real gas effects is for

Fig. 8: Isothermal areas in the symmetry line of a sphere,
M_∞ = 7.5km/sec, Kn_∞ = 0.08

correct simulations of reentry aerothermodynamics. The corresponding molecular and atomic concentrations along the stagnation streamline show the dissociation of O_2 and N_2 molecules in the shock front and the corresponding increase in atomic concentrations which lead to the formation of NO behind the shock front. The concentration distribution near the wall shows the recombination, which depends strongly upon the chosen boundary conditions.

3.3. Preliminary unsteady simulations

At reentry into the earth's atmosphere, the maximum heat transfer is expected at an altitude of approximately 70-80km. This can be physically understood because of the high speed which at this altitude is still approximately equal to the orbital speed, even though the atmosphere density has already increased strongly.

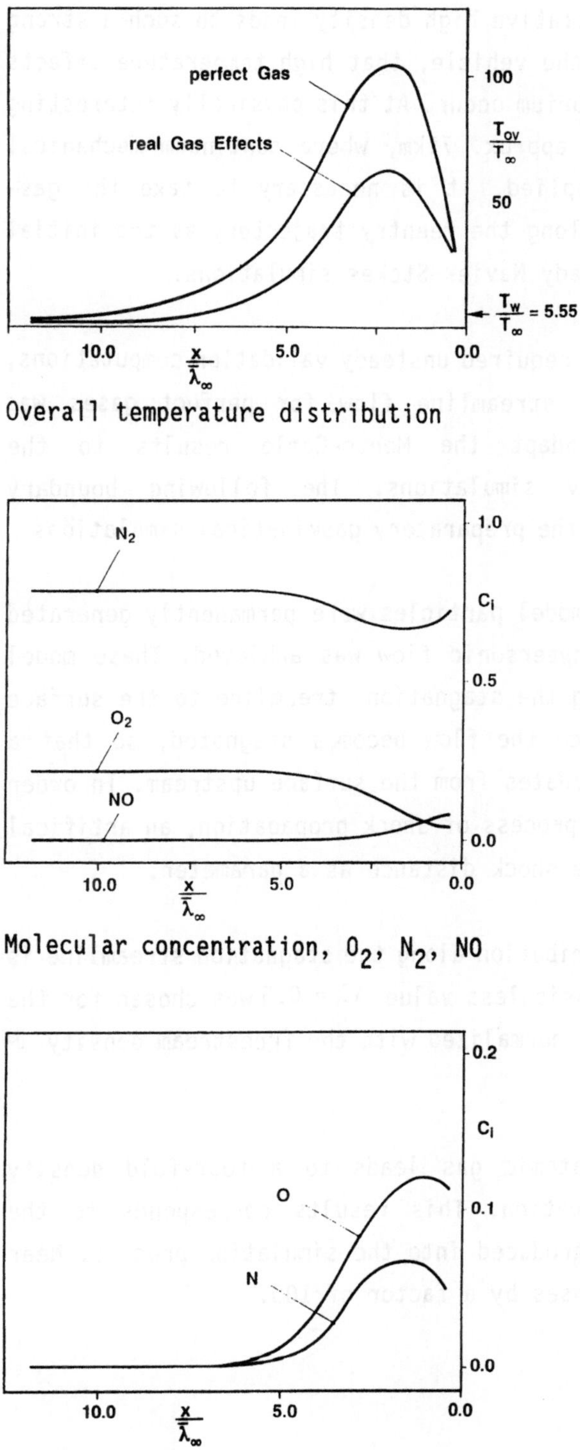

Fig. 9: Stagnationline, temperature and concentration profiles

The high velocity in the relative high density leads to such a strong heating of the gas around the vehicle, that high temperature effects in thermodynamic nonequilibrium occur. At this physically interesting and important altitude of approx. 75km, where continuum mechanical methods may already be applied, it is necessary to take the gas-kinetic results computed along the reentry trajectory as the initial distribution for the unsteady Navier-Stokes simulations.

As a first step towards the required unsteady validation computations, the stationary stagnation streamline flow for perfect gases was simulated in order to adapt the Monte-Carlo results to the requirements for unsteady simulations. The following boundary conditions were chosen for the preparatory gaskinetical simulations.

At the upstream boundary, model particles were permanently generated such that an undisturbed hypersonic flow was achieved. These model particles then travel along the stagnation streamline to the surface of the body. At the surface the flow becomes stagnated, so that a shock is formed which propagates from the surface upstream. In order to eliminate this unsteady process of shock propagation, an artifical outflow is allowed with the shock distance as a parameter.

In fig. 10 the density distribution along the stagnation streamline is illustrated. Here the dimensionless value $\Delta S = 0.5$ was chosen for the shock distance. The flow is normalized with the freestream density ρ_∞ and mean free path $\overline{\lambda_\infty}$.

The simulation of a monoatomic gas leads to a four-fold density increase at the shock location. This results corresponds to the assumptions, which were introduced into the simulation process. Near the wall the density increases by a factor of 100.

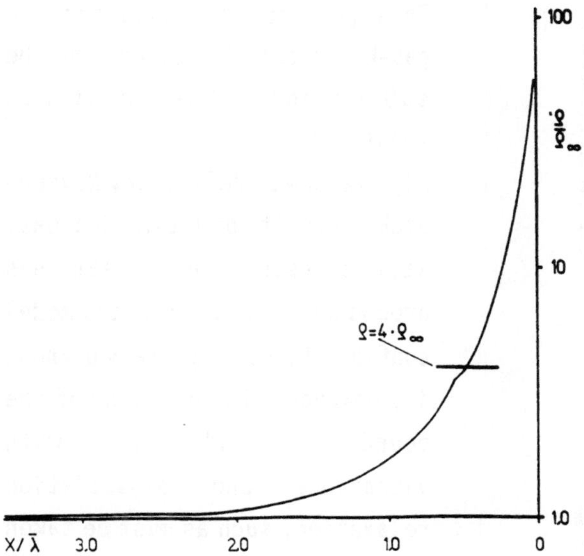

Fig. 10: Stagnation streamline density profile $M_\infty = 20$, $R = 1.67$, $T_W/T_\infty = 5$

4. NAVIER-STOKES SIMULATION

As numerous articles on Navier-Stokes simulations of hypersonic and reentry flows have already been written, this chapter will only offer a brief discussion of the problems. For a further treatment of the subject see [12] and [13], as no essentially new aspects have since become known. The problems here lie, as it was described in chapter 3, less in the numerical algorithms but much more in the lack of modelling of the reentry chemistry and the interaction with the body surface. We will return to this problem in chapter 5.

The first step towards a Navier-Stokes solution is the choice of a suitable grid. Fig. 11 shows a finite element grid, which, by means of an adapted refinement, respects the shock and the thermal boundary layer near the wall of the reentry vehicle.

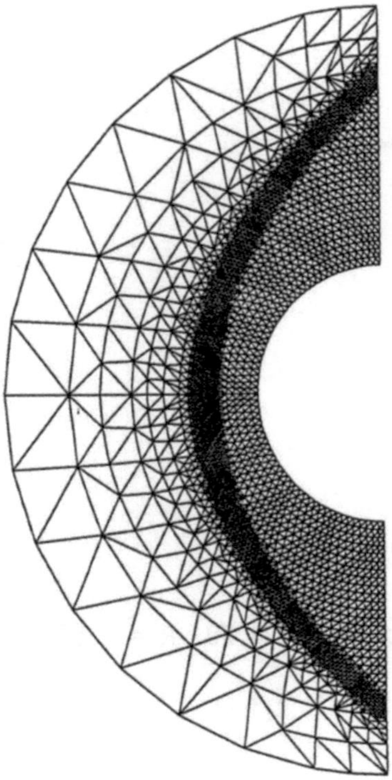

Fig. 11: Finite Element grid around a sphere

This grid is the basis for the gas-kinetical simulations and the subsequent Navier-Stokes solutions.

Fig. 12 shows preliminary Navier-Stokes results of Iso-Mach areas, wall stresses and streamlines around an ellipsoid reentry model configuration. Completely unknown is, however, the modelling of the boundary conditions with vibrations and dissociation relaxation, such as must be taken into account along a reentry trajectory. Since the conditions for the reentry of hot structures can not be simulated in wind tunnels, the definition of a quantitative free-flight reentry experiment is necessary here, too.

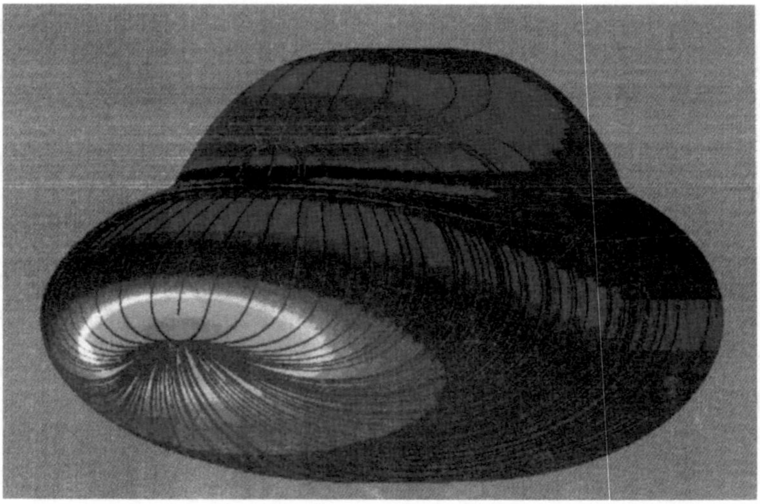

Fig. 12: Heat-flux and skin-friction lines around a double ellipsoid. RIEDELBAUCH, DLR Göttingen [13]

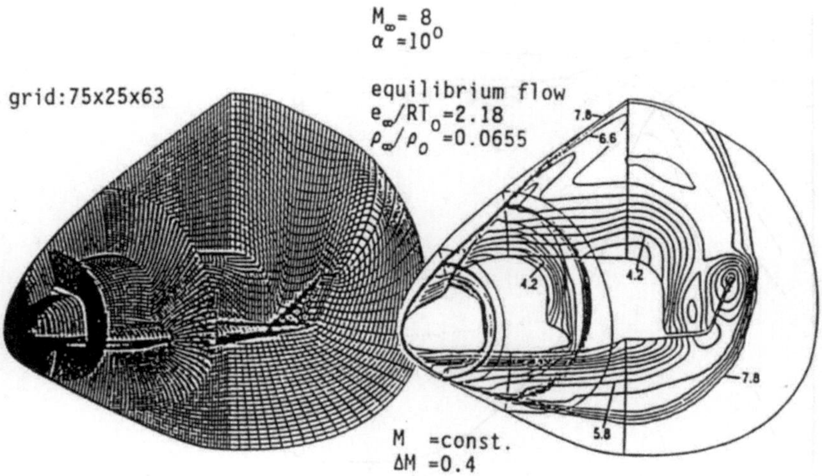

Fig. 13: Grid and iso-Machlines of an Euler solution around the HERMES configuration. RADESPIEL, DLR Braunschweig (1989)

In addition an Euler solution around the HERMES configuration is shown in fig. 13 which is from a physical or chemical point of view as good or physically insufficient as any Navier-Stokes solution which is available.

5. CHEMICAL MODELS

The key to the sucessful validation of numerical methods for reentry problems is the adequate chemical modelling of the chemical reactions in the high atmosphere. The situation today is such that numerical aerodynamics and numerical chemistry have developed seperately and with increasing sucess in solving their respective problems. Currently a practical common engineering approach is still lacking.

WARNATZ [14] shows in his most recent work, that the chemistry of air dissociation with five species leads to approximately 250 reactions, when rotation, vibration and dissociation interaction are taken into account.

In fig. 14, the mole fractions, X_i as functions of the temperature are depicted. The question then arises, as to which of the reactions

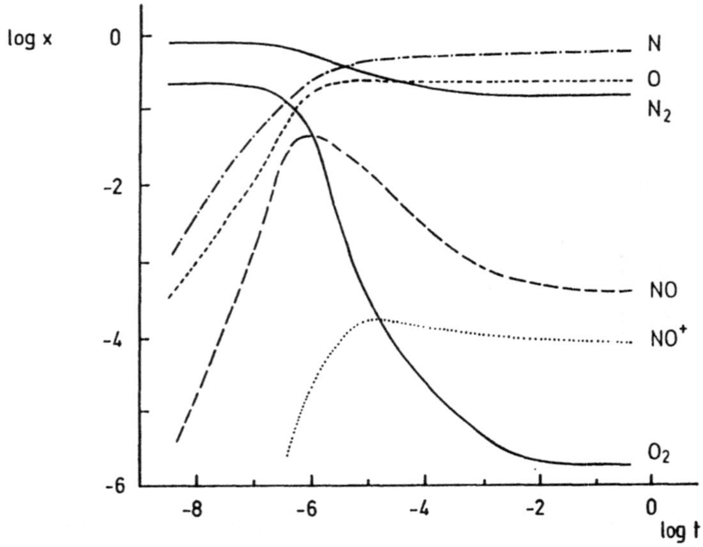

Fig. 14: Equilibrium mole fractions in air at an altitude of 71 km WARNATZ [14]

are really dominatly important with respect to the prediction of the heat flux in the heat shield. Here the numerical simulation methods need simple chemical models on the basis of modelling the energy and entropy fluxes in the flow field and on the surface.

6. CONCLUSION

The use of gas-kinetical and Navier-Stokes methods as design tools for reentry vehicles needs the intermediate step of validation. In the preceeding chapters we have argumented that a complete simulation of reentry aerothermodynamics requires that a quantitative free flight reentry experiment has to be carried out.

Presently in Germany, several reentry capsule concepts are being analysed. The University of Bremen's "BREMSAT" experiment is the first reentry experiment for the validation of gas-kinetical methods, and will be on board of the Space shuttle D2-mission planned for 1992.

The development of reentry flight techniques and the adaptation of numerical methods to the reentry trajectory, is pursued in the "COBRA

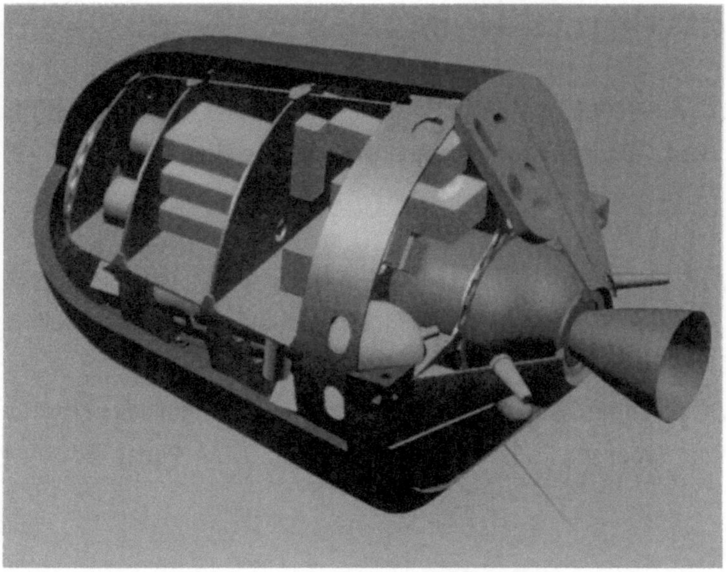

Fig. 15: Sketch of the RAUMKURIER and EXPRESS capsula
DASA: Dornier, MBB/ERNO (1989)

and EXPRESS" studies. The actual validation experiment is recently planned with EXPRESS capsula ahead of the HERMES and SÄNGER II projects. The necessary reentry technology program is worked out in [14].

Fig. 15 shows a sketch of the EXPRESS capsula. As described before, the adaption of the numerical simulation methods, along the reentry trajectory from the free molecular flow to continuum mechanics regime will be necessary for a successful validation. The validation experiment is presently in a definition stage.

LITERATURE

[1] DFG Memorandum: Rahmenkonzept für die Einrichtung von Sonderforschungsbereichen zur Erforschung von Grundlagen für den Hyperschallflug, (1988).

[2] Oertel Jr., H.: Nutzung von Vektorrechnern für die numerische Aerodynamik. PIK 11, 3, 164-170 (1988).

[3] Bird, G.A.: Direct Molecular Simulation of a Dissociating Diatomic Gas. Journal of Computational Physics, 25, 353-365 (1977).

[4] Bird, G.A.: Low Density Aerothermodynamics. AIAA Paper 85-0994 (1985).

[5] Moss, J.N.; Cuda Jr., V.; Simmonds, A.L.: Nonequilibrium Effects for Hypersonic Transitional Flows. AIAA Paper 87-0404 (1987).

[6] Wetzel, W.: Gaskinetische Simulation von Hyperschallströmungen um dreidimensionale stumpfe Körper unter Wiedereintrittsbedingungen, ZLR-Forschungsbericht, TU Braunschweig, 89-02, (1989).

[7] Wetzel, W.; Oertel Jr., H.: Gas-Kinetical Simulation of Vortical Flows. Acta Mechanica, 70, 127-143 (1987).

[8] Borgnakke, C.; Larsen, P.S.: Statistical Collision Model for Monte-Carlo-Simulation of Polyatomic Gas Mixtures. Journal of Computational Physics, 18, 405-420 (1975).

[9] Russel, D.A.: Density Disturbance ahead of a Sphere in Rarefied Supersonic Flow. The Physics of Fluids, 11, (1968).

[10] Bortner, M.H.: A Review of Rate Constants of Selected Reactions of Interest in Reentry Flow Fields in the Atmosphere. National Bureau of Standards, Technical Note 484 (1969).

[11] Larsen, P.S.: Monte Carlo Simulation of a Homogeneous Dissociating Diatomic gas. Archiv of Mech., 28, 719 (1976).

[12] Oertel Jr., H.: Validation of Navier-Stokes Methods for Aerospace Applications. Proc. of the International Symposium on Computational Fluid Dynamics, Nagoya (1989)

[13] Riedelbauch, S.; Wetzel, W.; Kordulla, W.; Oertel Jr. , H.: Proceedings of the 16th International Sypmosium of Rarified Gasdynamics 1988: On the Numerical Simulation of Three-Dimensional Hypersonic Flow, AGARD CP-428, (1988).

[14] Warnatz, J.: Different Levels of Air Dissociation Chemistry and its Coupling with Flow Models. The Second Joint Europe/US - Short Course in Hypersonics. Colorado Springs (1989)

[15] Oertel Jr. H.: Wiedereintrittstechnologisches Experimentalprogramm, BMFT 01 QV 8934 (1990)

Numerical Simulation of Boltzmann Flows of Real Gases
– Accuracy of Models Used in the Monte Carlo Method –

K. Nanbu

Institute of Fluid Science, Tohoku University,
Katahira, Aoba, Sendai 980, JAPAN

Accuracy of some models used in the Monte Carlo direct simulation method is examined by comparing the Monte Carlo solutions at small Knudsen numbers with the exact solutions of the Navier-Stokes equations. The test flow is a hypersonic Couette flow. The variable-hard-sphere model is accurate. The use of the Larsen-Borgnakke model for translational-rotational energy transfer causes 4 % error. If this model is extended so as to include vibrational energy transfer, the error reaches 7 %. Lastly, a simple method for simulating dissociating-recombinating flows is proposed.

1. Introduction

The Monte Carlo direct simulation method (MCDS) has a strict mathematical foundation if it is used for computation of monatomic gas flows; the Monte Carlo solution is surely a solution of the Boltzmann or Kac-Leontovich equation.[1] Of course, a proper choice of cell size and time step must be done even in this case. When we want to compute diatomic gas flows by the MCDS method, some model for internal energy transfer should be introduced. The Larsen-Borgnakke (LB) model is usually used in translatinal-rotational (T-R) energy transfer.[2] This is extended to the case of translational-rotational- vibrational (T-R-V) energy transfer.[3] The validity of the model is checked by comparing computed solutions with limited experimental data.[4] Another way of checking is to compare the Monte Carlo solution at a very small Knudsen number with the exact, if possible, solution of the Navier-Stokes equation. Here, accuracy of the variable-hard-sphere (VHS) model for a monatomic gas is first examined. Then, accuracy of the LB model for T-R and T-R-V energy transfers is examined along the latter line. The test flow is a hypersonic Couette flow between cold and very hot walls. In the last, a simple method for calculating dissociating-recombinating flow is proposed together with a new recombination model.

2. Exact Navier-Stokes Solutions

The exact solution of the Navier-Stokes (NS) equation is a touchstone for the models used in the MCDS method. Here we consider the Couette flow under the following conditions. The hot wall at $y = 0$ with temperature T_w is at rest and the cold wall at $y = h$ with temperature T_∞ is moving with a constant speed U. Velocity slip and temperature jump are neglected. To be specific, we choose

$$T_w = 1500 \text{ K}, \quad T_\infty = 300 \text{ K}, \quad U = 7 \text{ km/s}. \tag{1}$$

Gas is a pure oxygen. The viscosity μ is proportional to gas temperature T. The Prandtl number is constant and is given by Eucken's relation

$$Pr = 4\gamma/(9\gamma - 5), \tag{2}$$

where γ is the specific heat ratio. The exact solutions can be obtained for two cases.[5]

2.1 Constant γ

This is the case of a monatomic gas or a diatomic gas with only rotational internal energy. The ratio of heat flux q_w (from fluid to hot wall) to shear stress τ_w is

$$\tilde{q}_w \equiv \frac{q_w}{\tau_w U} = \frac{c_p(T_r - T_w)}{PrU^2}, \tag{3}$$

where $T_r/T_\infty = 1 + A$, $A = 0.5Pr(\gamma - 1)M_\infty^2$, and $M_\infty = U/(\gamma R T_\infty)^{1/2}$, R being the gas constant for oxygen. The shear stress is

$$\tilde{\tau}_w \equiv \frac{\tau_w h}{\mu_\infty U} = 1 - A\tilde{q}_w + \frac{2}{3}A. \tag{4}$$

The velocity $u/U(= \xi)$ is given in the implicit form

$$\xi - A\tilde{q}_w(2\xi - \xi^2) + A(\xi - \xi^3/3) = \tilde{\tau}_w y/h. \tag{5}$$

Once y/h is given, ξ can be easily obtained by the inverse interpolation. The temperature T/T_∞ is

$$T/T_\infty = 1 - 2A\tilde{q}_w(1 - \xi) + A(1 - \xi^2). \tag{6}$$

2.2 Diatomic gas with vibrational energy

The specific heat c_v is

$$c_v/R = 2.5 + (\eta/\sinh\eta)^2, \tag{7}$$

where $\eta = \theta_v/2T$ and θ_v is the characteristic temperature for vibration. It is 2270 K for oxygen. Using eq.(7), we have

$$\tilde{q}_w = \frac{1}{2} - \frac{R\theta_v}{2PrU^2}[F(\eta_w) - F(\eta_\infty)], \tag{8}$$

where $F(\eta) = 3.5/\eta + \coth\eta$, $\eta_w = \theta_v/2T_w$, and $\eta_\infty = \theta_v/2T_\infty$. The relation between velocity and temperature is

$$\frac{1}{2}\xi^2 - \tilde{q}_w\xi = \frac{R\theta_v}{2PrU^2}[F(\eta_w) - F(\eta)]. \tag{9}$$

From eq.(9) we have $\eta(\xi)$. Then,

$$\tilde{\tau}_w = \eta_\infty \int_0^1 [\eta(\xi)]^{-1} d\xi. \tag{10}$$

The velocity profile u/U is calculated from

$$\int_0^{u/U} [\eta(\xi)]^{-1} d\xi = \frac{\tilde{\tau}_w y}{\eta_\infty h}. \tag{11}$$

Let us compare the Monte Carlo solution of $\tilde{\tau}_w$, $\tilde{q}_w$, u/U, and T/T_∞ with the NS solution.

3. Monatomic Gas

The number of binary collisions in $(t, t+\Delta t)$ is a random variable. However, if $g\sigma_T(g)$ is independent of g, the average collision number is constant and known at time t. Here g is the relative velocity and $\sigma_T(g)$ is the total collision cross-section. Let us use the average collision number in $(t, t+\Delta t)$. This is the collision-number scheme.[6] Since the viscosity μ is assumed to be proportional to T, our molecular model is Maxwellian. In this case, $g\sigma_T(g)$ is really constant. But the collision-number scheme can be used for general inverse-power molecules if the idea of maximum collision frequency is introduced.[7,8]

3.1 Exact Monte Carlo scheme

First we examine the accuracy of the MCDS method itself. The collision probability in $(t, t+\Delta t)$ of a simulator molecule is

$$P_c = \pi\beta_0^2\left(\frac{8a}{m}\right)^{1/2}\frac{N-1}{N}n\Delta t, \tag{12}$$

where N is the number of simulators in a cell, m is the mass of a molecule, a is the potential constant, and β_0 is the cut-off value of dimensionless impact parameter $\beta[= b(mg^2/8a)^{1/4}]$. If the cut-off of deflection angle is 2°, then $\beta_0 = 2.391$. Note that $P_c/(N-1)$ is the collision probability of a pair of simulators. Since this does not depend on g, the pair can be chosen at random. The collision number N_c is given by

$$N_c = NP_c/2 = \pi\beta_0^2(2a/m)^{1/2}(N-1)n\Delta t. \tag{13}$$

The time step Δt has been so chosen that the condition $P_c < 0.5$ is satisfied.

Momentum and energy are conserved in each collision in the collision-number scheme. The deflection angle χ is given by[9]

$$\chi = \pi - 3.1182\beta(1 + \beta^2(-0.1618 + \beta^2(-0.0615 + 0.0352\beta^2))), (\beta \leq 1)$$

$$\chi = 1.1781\beta'(1 + \beta'(-1.0770 + \beta'(1.2460 + \beta'(-0.9975 + 0.3472\beta')))), (\beta > 1)$$

where $\beta' = \beta^{-4}$ and $\beta = \beta_0 R^{*1/2}$, R^* being a random fraction. The Knudsen number Kn is λ/h. The mean free path λ is defined by

$$\lambda = \frac{\mu_w}{\rho_0}\left(\frac{\pi}{2RT_w}\right)^{1/2}, \tag{14}$$

where ρ_0 is the average density between the two walls. Note that ρ_0 is independent of time in case of the Couette flow. The viscosity μ_w is

$$\mu_w = \frac{kT_w}{3\pi A_2}\left(\frac{m}{2a}\right)^{1/2},$$

where $A_2 = 0.4362$ and k is the Boltzmann constant. The values of Kn are chosen as

$$Kn = 0.01, 0.02, 0.03, 0.05, 0.07, 0.10.$$

The initial condition is

$$u = 0, \quad T = (T_w + T_\infty)/2.$$

The time at which the flow becomes steady is determined by examining the temporal behavior of τ_w and q_w. The boundary condition is a diffuse reflection with complete thermal accomodation.

We have found that the size and shape of cell have a large effect on the temperature profile. Figure 1, which includes only a part of the Monte Carlo

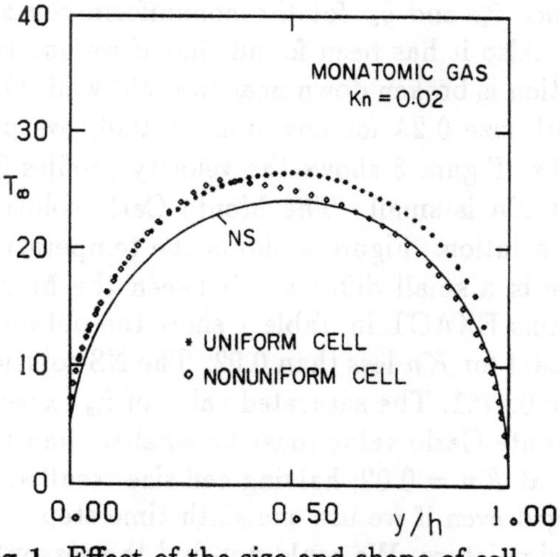

Fig.1 Effect of the size and shape of cell.

Fig.2 Nonuniform cell.

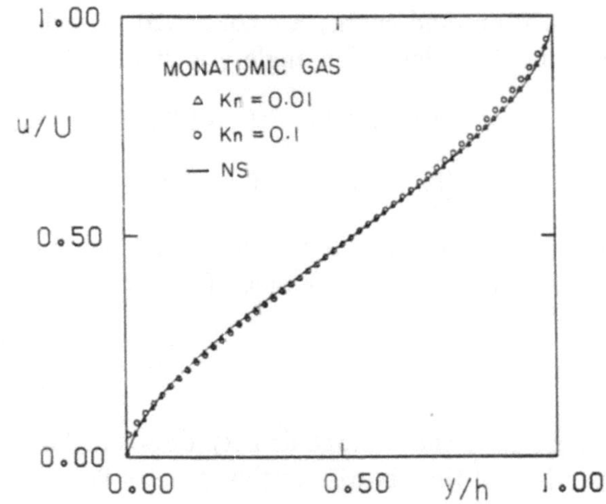

Fig.3 Flow velocity (exact Monte Carlo).

data, shows solutions for two cases; one is 250 cells with a uniform size and the other is 100 cells whose size is, as shown in Fig.2, smaller near the walls where the gradients of T and u are large. In the latter case the coordinates y_i of the cell boundary are given by[10]

$$\frac{y_j}{h} = \frac{(a+1)b - a + 1}{2(1+b)},$$

where

$$b = \left(\frac{a+1}{a-1}\right)^{2y^*-1},$$

$y^* = j/J(j = 0, 1, \cdots, 100)$, and $a = 1.08$. The result for the nonuniform cell must be wrong because the temperature near the cold wall at $y/h = 1$ is lower than the NS solution. However, the temperatures for the two types of cell agree near the hot wall, hence $\tilde{\tau}_w$ and $\tilde{q}_w$ for the nonuniform cell are equal to those for the uniform cell. Also it has been found that if we use 100 uniform cells, the Monte Carlo solution is broken down near the cold wall. Our final choice is the uniform cells with size 0.2λ for any Kn. Initially we put 30 simulators into each of these cells. Figure 3 shows the velocity profiles for $Kn = 0.01$ and 0.1. The effect of Kn is small. The Monte Carlo solution for $Kn = 0.01$ agrees with the NS solution. Figure 4 shows the temperature profiles. Even at $Kn = 0.01$ there is a small difference between the Monte Carlo and NS solutions. The columns EXACT in Table 1 show the obtained $\tilde{\tau}_w$ and $\tilde{q}_w$. The shear stress is constant for Kn less than 0.02. The NS solution gives $(\tilde{\tau}_w)_{NS} = 16.97$ and $(\tilde{q}_w)_{NS} = 0.4761$. The saturated value of $\tilde{\tau}_w$ exceeds the NS value by 2.6 %.Since the Monte Carlo value must be smaller than the NS one, we searched for the reason at $Kn = 0.02$; halving cell size resulted in 2.1 % overshoot and it is not changed even if we use one sixth time step, half cut-off angle, or double number of simulators. We could not find the reason of

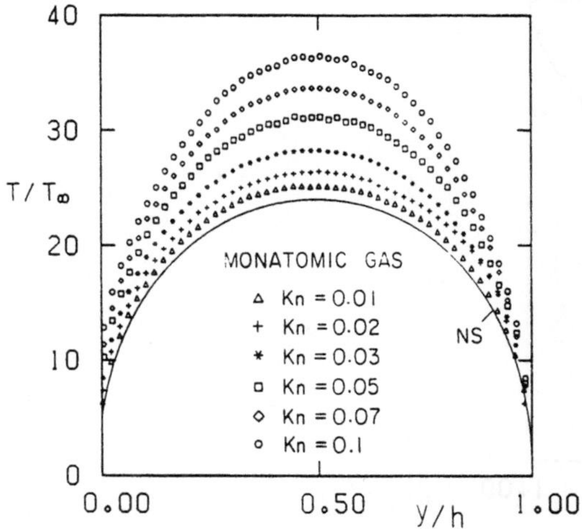

Fig.4 Temperature (exact Monte Carlo).

Table 1 Shear stress $\tilde{\tau}_w$ and heat-flux to shear-stress ratio $\tilde{q}_w$

	$\tilde{\tau}_w$		$\tilde{q}_w$	
Kn	EXACT	VHS	EXACT	VHS
0.10	11.94	12.05	0.4756	0.4783
0.07	14.05	14.08	0.4749	0.4772
0.05	15.75	15.60	0.4749	0.4759
0.03	17.04	17.02	0.4743	0.4764
0.02	17.41	17.55	0.4743	0.4747
0.01	17.40	17.42	0.4762	0.4808

the overshoot. We think that this small error is an essential one to the MCDS method, because the collision partner of a molecule near a cell boundary is chosen only from a cell on one side.

3.2 Variable hard sphere model

Here we examine the accuracy of the VHS model. The total cross-section is given by

$$\sigma_T = \sigma_w (6RT_w)^{1/2}/g, \tag{15}$$

where

$$\sigma_w = (2kmT_w/3)^{1/2}/\mu_w. \tag{16}$$

The collision probability P_c is

$$P_c = \sigma_w (6RT_w)^{1/2}[(N-1)/N]n\Delta t. \tag{17}$$

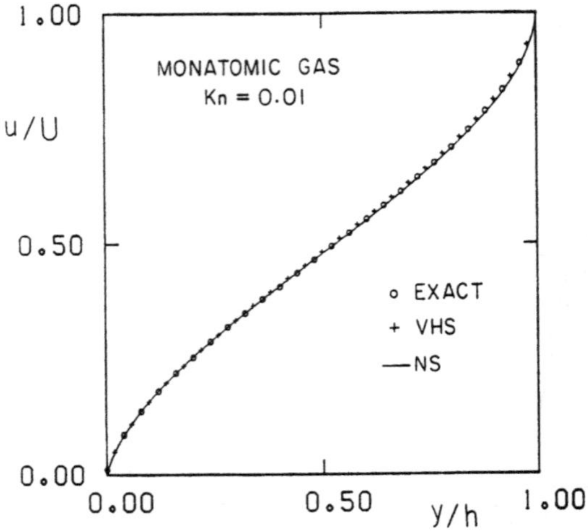

Fig.5 Flow velocity. Comparison of the exact Monte Carlo, Variable-Hard-Sphere, and Navier-Stokes solutions.

The collision number N_c is

$$N_c = (1/2)\sigma_w(6RT_w)^{1/2}(N-1)n\Delta t. \tag{18}$$

To compare the solution obtained by use of this model with the exact Monte Carlo solution, we do not introduce the VHS definition of mean free path; eq.(14) is used. Of course, the postcollision velocities follow the scattering law of hard sphere.

The exact and VHS solutions at $Kn = 0.01$ are compared in Figs.5 and 6. The agreement is good. This is true also for other Knudsen numbers. The shear stress and heat flux are compared in Table 1. The use of the VHS model causes no significant errors. We use the VHS model hereafter.

4. Diatomic Gas with Rotational Energy

Accuracy of the LB model for T-R energy transfer can be examined by comparing the Monte Carlo solution with the NS solution for $\gamma = 7/5$. The latter gives $(\tilde{\tau}_w)_{NS} = 14.03$ and $(\tilde{q}_w)_{NS} = 0.4698$. Let P_{EL} and P_{TR} be the probabilities of elastic and inelastic collisions, respectively. Then we have

$$P_{EL} + P_{TR} = 1.$$

Here P_{TR} is chosen in such a way that the difference between the Monte Carlo and NS solutions takes a minimum.[11]

Although the post-collision velocities follow the scattering law of the hard sphere in using the VHS model, the LB model must be used within the framework of Maxwellian molecule, which is the present molecular model. Let ε_i be the rotational energy of molecule i before collision. The total collision energy

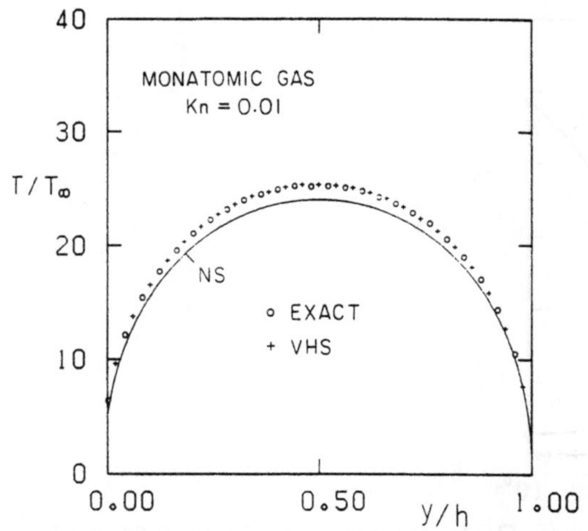

Fig.6 Temperature. Comparison of the exact Monte Carlo, Variable-Hard-Sphere, and Navier-Stokes solutions.

F_c in the center-of-mass system is

$$F_c = E_T + \varepsilon_1 + \varepsilon_2 = E'_T + \varepsilon'_1 + \varepsilon'_2, \tag{19}$$

where $E_T (= mg^2/4)$ is the kinetic energy of a collision pair, $g = |\mathbf{c}_2 - \mathbf{c}_1|$, and primes denote the post-collision state. We can find samples of $\mathbf{c}'_1$, $\mathbf{c}'_2$, ε'_1, and ε'_2 from the LB model. Let X be E'_T/F_c. Then the probability density function normalized by its maximum is

$$f(X) = (3\sqrt{3}/2)\sqrt{X}(1 - X). \tag{20}$$

Once X is sampled by the rejection method, we have

$$E'_T = \frac{1}{4}mg'^2 = XF_c. \tag{21}$$

From eq.(19),

$$\varepsilon'_I \equiv \varepsilon'_1 + \varepsilon'_2 = F_c - E'_T. \tag{22}$$

The energies ε'_1 and ε'_2 are

$$\varepsilon'_1 = R^* \varepsilon'_I, \quad \varepsilon'_2 = \varepsilon'_I - \varepsilon'_1. \tag{23}$$

Since g' can be obtained from eq.(21), we have

$$\mathbf{g}' = \mathbf{c}'_2 - \mathbf{c}'_1 = g'\mathbf{R}^*, \tag{24}$$

where $\mathbf{R}^*$ is the unit random vector. The conservation of momentum is

$$\mathbf{c}'_1 + \mathbf{c}'_2 = \mathbf{c}_1 + \mathbf{c}_2. \tag{25}$$

The post-collision velocities $\mathbf{c}'_1$, $\mathbf{c}'_2$ can be obtained from eqs. (24) and (25).

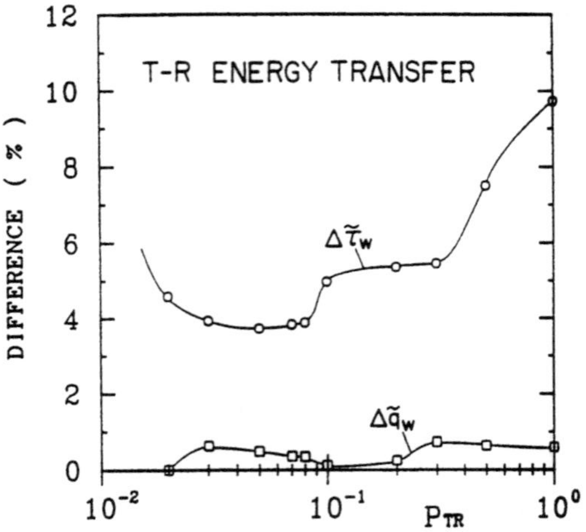

Fig.7 Difference between the Monte Carlo solution at $Kn = 0.01$ and Navier-Stokes solution (T-R energy transfer).

Let $\tilde{\tau}_w$ and $\tilde{q}_w$ be the Monte Carlo solution at $Kn = 0.01$. Figure 7 shows $\Delta\tilde{\tau}_w$ and $\Delta\tilde{q}_w$ as a function of P_{TR}, where $\Delta\tilde{\tau}_w = [\tilde{\tau}_w - (\tilde{\tau}_w)_{NS}]/(\tilde{\tau}_w)_{NS}$, $\Delta\tilde{q}_w$ being similarly defined. It is seen that $\Delta\tilde{\tau}_w$ takes a minimum of about 4 % at $P_{TR} = 0.05$. Figure 8 shows the flow velocity and temperature for $Kn = 0.01$ and $P_{TR} = 0.05$. As for the velocity, the Monte Carlo solution agrees with the NS solution, whereas there is a small difference between the two temperatures.

5. Diatomic Gas with Rotational and Vibrational Energy

The LB model is extended to the T-R-V energy transfer.[3] Before using the extended model, we must assign the probability of inelastic collision that the T-R-V energy transfer occurs. Of course, we have

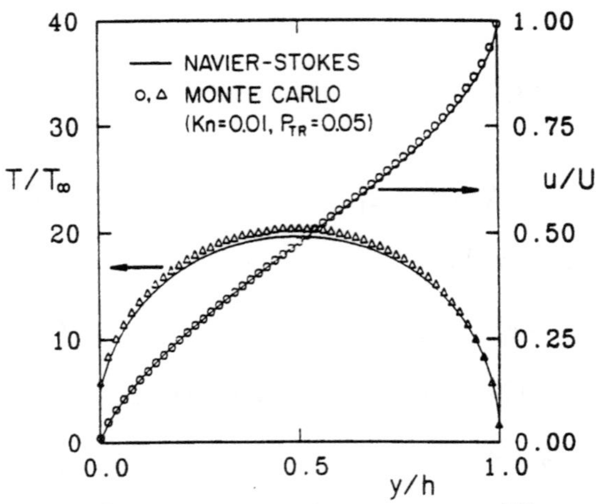

Fig.8 Flow velocity and temperature ($Kn = 0.01$, $P_{TR} = 0.05$).

$$P_{EL} + P_{TR} + P_{TRV} = 1. \tag{26}$$

Since the LB model is not a collision-dynamical but a statistical model, inelastic collision probabilities have no direct connection with measured probabilities, which are the outcome of collision dynamics of real molecules. We soon introduce a simple assumption for P_{EL}, P_{TR}, and P_{TRV} in connection with the LB model.

Let ε_i be the internal energy of molecule i before collision. It is

$$\varepsilon_i = \varepsilon_i^R + \varepsilon_i^V,$$

where ε_i^R and ε_i^V denote the rotational and vibrational energies. The total collision energy E_c is

$$E_c = E_T + \varepsilon_1 + \varepsilon_2 = E_T' + \varepsilon_1' + \varepsilon_2'. \tag{27}$$

Note that eq.(27) is different from eq.(19) because ε_i is the sum of ε_i^R and ε_i^V. Our purpose is to determine

$$\mathbf{c}_i', \varepsilon_i^{R'}, \varepsilon_i^{V'} (i = 1, 2).$$

The essential assumption of the LB model is that these post-collision quantities are distributed according to an equilibrium. The equilibrium temperature T_c is determined from

$$\langle E_T' \rangle + \langle \varepsilon_1' \rangle + \langle \varepsilon_2' \rangle = E_c, \tag{28}$$

where $\langle\ \rangle$ denotes the average in equilibrium, i.e.,

$$\langle E_T' \rangle = \frac{3}{2} kT_c, \tag{29}$$

$$\langle \varepsilon_i' \rangle = kT_c + \frac{k\theta_v}{\exp(\theta_v/T_c) - 1}, \quad (i = 1, 2). \tag{30}$$

Note that eq.(29) holds for Maxwellian molecule. (In case of hard sphere, 3/2 must be replaced by 2.) Substitution of eqs.(29) and (30) into eq.(28) yields

$$F(x) \equiv \frac{7}{4} x + \left(\exp \frac{1}{x} - 1\right)^{-1} - \tilde{E}_c = 0, \tag{31}$$

where $x = T_c/\theta_v$, and $\tilde{E}_c = E_c/2k\theta_v$. The solution is

$$x = (4/7)\tilde{E}_c \quad \text{for} \quad \tilde{E}_c < 0.2.$$

In other case we use Newton's method. The first approximation $x^{(1)}$ is

$$x^{(1)} = (4/7)\tilde{E}_c \quad \text{for} \quad \tilde{E}_c \leq 1.8,$$

$$x^{(1)} = (4/11)\tilde{E}_c \quad \text{for} \quad \tilde{E}_c > 1.8.$$

The $(k+1)$-th approximation is given by

$$x^{(k+1)} = x^{(k)} - [F(x^{(k)})/F'(x^{(k)})]$$

We have used the convergence criterion of

$$|\, 1 - x^{(k)}/x^{(k+1)} \,| < 0.001.$$

Now let us introduce a model for P_{EL}, P_{TR}, and P_{TRV}. Note that T_c increases with E_c. The probability P_{TRV} is expected to be small for low T_c. Taking this into consideration, we propose the following model[11]

$$P_{TR} = \alpha(2/\zeta_0), \;\; P_{TRV} = \alpha(\zeta_v/\zeta_0), \tag{32}$$

where α is an adjustable constant, $\zeta_0 = 5+\zeta_v$, and ζ_v is the apparent vibrational degree of freedom defined by

$$\zeta_v = (2\theta_v/T_c)/[\exp(\theta_v/T_c) - 1]. \tag{33}$$

This equation means that the probability of an inelastic collision is proportional to the corresponding internal energy.

There are three possible events in collision. Which occurs can be simulated by calling a random fraction. If the T-R event occurs, we follow the procedure in §4; the vibrational energies ε_1^V, ε_2^V do not change. Here we consider the T-R-V event. Let X be E_T'/E_c. The probability density of X normalized by its maximum is

$$\sqrt{2\zeta - 1}[(\zeta - 0.5)/(\zeta - 1)]^{\zeta-1}\sqrt{X}(1 - X)^{(\zeta-1)},$$

where $\zeta = 2 + \zeta_v$. Once X is sampled, E_T' is given by

$$E_T' = mg'^2/4 = XE_c. \tag{34}$$

The sum of internal energies is

$$E_I' \equiv \varepsilon_1' + \varepsilon_2' = E_c - E_T'.$$

Let $Y = \varepsilon_1'/E_I'$. The probability density of Y normalized by its maximum is

$$2^{\zeta-2}[Y(1 - Y)]^{0.5\zeta-1}.$$

If Y is sampled, we have

$$\varepsilon_1' = YE_I', \;\; \varepsilon_2' = E_I' - \varepsilon_1'.$$

The energies ε_1', ε_2' are divided into two modes in proportion to the degree of freedom

$$\varepsilon_i^{R'} = (2/\zeta)\varepsilon_i', \;\; \varepsilon_i^{V'} = \varepsilon_i' - \varepsilon_i^{R'}, \quad (i = 1, 2).$$

The post-collision velocities $\mathbf{c}_1'$, $\mathbf{c}_2'$ can be obtained from eqs.(24) and (25), where g' is given by eq.(34).

The NS solution gives $(\tilde{\tau}_w)_{NS} = 11.88$ and $(\tilde{q}_w)_{NS} = 0.4652$. The Monte Carlo solution is obtained for $Kn = 0.01$. Figure 9 shows $\Delta\tilde{\tau}_w$ and $\Delta\tilde{q}_w$ as a

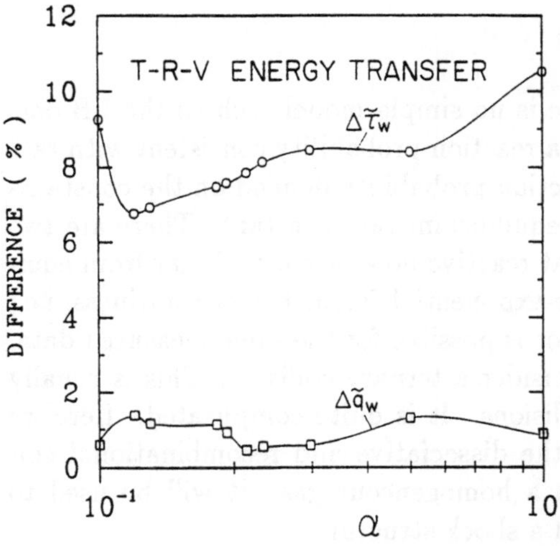

Fig.9 Difference between the Monte Carlo solution at $Kn = 0.01$ and Navier-Stokes solution (T-R-V energy transfer).

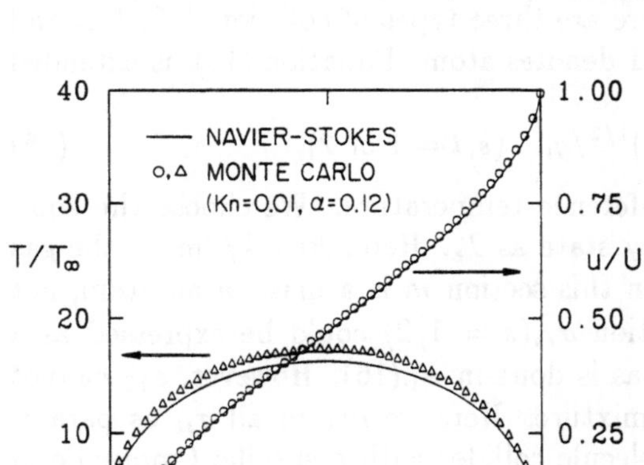

Fig.10 Flow velocity and temperature ($Kn = 0.01$, $\alpha = 0.12$).

function of α. We see that $\Delta\tilde{\tau}_w$ takes a minimum of about 7 % at $\alpha = 0.12$. Note that for this value of α, eq.(32) gives $P_{TR} = 0.048$ if ζ_v were zero. This value is chose to 0.05 that yields the minimum $\Delta\tilde{\tau}_w$ in Fig.7. The Monte Carlo solution for $\alpha = 0.12$ is compared with the NS solution in Fig.10. The flow velocities for the two solutions agree well. The overall temperature T is obtained from

$$\frac{5}{2}kT + \frac{k\theta_v}{\exp(\theta_v/T) - 1} = \frac{1}{2}m(\bar{c^2} - u^2) + \bar{\varepsilon}^R + \bar{\varepsilon}^V,$$

where the bars denote average and c, ε^R, and ε^V are the speed, rotational energy, and vibrational energy of a molecule. Differently from u, the temperature at $Kn = 0.01$ shows a small deviation from that of the NS solution.

6 . Dissociation and Recombination

In case of reactive collision,there is no simple model such as the LB one. As for dissociation, Bird[12] proposed a reaction probability consistent with rate constant. The parameters in the reaction probability depend on the constants in the pre-exponential factor of the equilibrium rate constant. There are two problems in this approach; one is that reactive flows are usually far from equilibrium and the other is that the pre-exponential factor has arbitrariness, i.e., a quite different pre-exponential factor is possible for the same measured data. As for recombination, we have to consider a ternary collision. This is usually treated as two successive binary collisions. It is quite complicated. Here we propose a much simpler model for the dissociative and recombinational collision. We apply our model only to a homogeneous gas. It will be used to calculate a reactive Couette flow and a shock structure.

As before, the molecular model is Maxwellian molecule treated by the VHS model. Also, the collision number scheme is used. The gas is a mixture of molecular and atomic oxygen. There are three types of collision, 2-2, 1-1, and 2-1, where 2 denotes molecule and 1 denotes atom. Equation (15) is extended as follows.

$$\sigma_T^{st} = \sigma_{st}(6RT_e)^{1/2}/g, \quad (s,t = 1 \text{ or } 2), \tag{35}$$

where $\sigma_{12} = \sigma_{21}$, and T_e is the reference temperature. We choose the equilibrium temperature at a stationary state as T_e. Here, $R(= k/2m)$ is the gas constant for molecule. Note that in this section m is a mass of an atom, not molecule. The reference cross-section $\sigma_{ss}(s = 1,2)$ could be expressed as a function of viscosity of a pure gas as is done in eq.(16). However, σ_{12} cannot be specified from the viscosity of mixture. Here we regard all σ_{st} as parameters. The probability that a s-molecule collides with a specific t-molecule in $(t, t+\Delta t)$is[13]

$$P_{st} = \frac{n_t \Delta t}{N_t} g\sigma_T^{st} = \frac{1}{N_t} n_t \Delta t \sigma_{st}(6RT_e)^{1/2}, \tag{36}$$

where N_t is the number of simulator molecules of species t. The collision number between like molecules is

$$N_{ss} = \frac{1}{2}N_s(N_s - 1)P_{ss} = \frac{1}{2}(N_s - 1)n_s \Delta t \sigma_{ss}(6RT_e)^{1/2}, \tag{37}$$

where $s = 1$ or 2. The collision number between unlike molecules is

$$N_{12} = N_1 N_2 P_{12} = N_1 n_2 \Delta t \sigma_{12}(6RT_e).^{1/2} \tag{38}$$

Dissociation occurs in 2-2 or 1-2 collision whereas recombnination does only in 1-1 collision. The time step Δt should be chosen in such a way that the collision probability of a particle is smaller than unity, i.e., $(N_2 - 1)P_{22}$, $(N_1 - 1)P_{11}$, $N_2\ P_{12}$, and $N_1\ P_{21}$ are less than unity.

6.1 Structure of collisional events

If dissociation is forbidden, possible events in a 2-2 collision are elastic, T-R, and T-R-V ones. These probabilities satisfy

$$P_{EL} + P_{TR} + P_{TRV} = 1. \tag{39}$$

The post-collision quantities c'_i, $\varepsilon_i^{R'}$,$\varepsilon_i^{V'}$ $(i = 1, 2)$ can be determined by the procedure in §5. Our dissociation model is as follows. A molecule ($i = 1$ or 2) dissociates if

$$\varepsilon_i^{V'} + \delta\varepsilon_i^{R'} > D \tag{40}$$

is satisfied, where D is the dissociation energy and δ, which is between 0 and 1, is a free parameter. First all 2-2 collisions are calculated N_{22} times in this way. At the cend of this stage, N_1 and N_2 are renewed. Using these renewed numbers, we obtain N_{12} from eq.(38).

Secondly, we calculate 1-2 collisions N_{12} times without allowing dissociation. A 1-2 collision results in elastic, T-R, or T-R-V event. How to determine the post-collsion quantities is explained in §6.2. Dissociation occurs if the condition (40) is satisfied for molecule. After all 1-2 collisions are calculated, we obtain N_{11} by using the last value of N_1.

Lastly, we calculate 1-1 collisions N_{11} times. A 1-1 collision results in elastic collision or recombination. That is, if a pair of atoms can find neighboring third particle (atom or molecule) which acts as an absorber of the dissociation energy, the pair turns into a molecule. We propose the following procedure for the calculation of recombination. During the computation of all 2-2 and 1-2 collisions, the number of dissociated molecules is counted. Let M_D be the total number of such molecules in one time step and N_1 and N_2 be the number of atoms and the number of molecules at the end of all 2-2 and 1-2 collisions. The degree of dissociation is then

$$\alpha = N_1/(2N_2 + N_1). \tag{41}$$

We assume that the number of recombinational collisions is given by

$$M_R = (\alpha/\alpha_e)^2 M_D, \tag{42}$$

where α_e is the degree of dissociation in a local equilibrium. The number M_R is equal to M_D if $\alpha = \alpha_e$ and it is proportional to the number of 1-1 collisions N_{11} if $N_1(N_1 - 1) \simeq {N_1}^2$. Once M_R is given, the number of elastic 1-1 collisions is $N_{11} - M_R$. The value of α_e is constant for a homogenous gas in an isolated system. In case of nonuniform gases it changes with time and position in flow field. Here we describe how to determine α_e in the general case. The Lighthill formula is

$$\frac{{\alpha_e}^2}{1-\alpha_e} = \frac{\rho_D}{\rho}\exp\Big(-\frac{\theta_D}{T_e}\Big), \tag{43}$$

where ρ_D and θ_D are characteristic density and temperature for dissociation; $\rho_D = 150$ g/cm^3 and $\theta_D = 59500$ K for O_2. Let us give our attention to a cell in flow field. There are atoms and molecules in the cell. The density ρ

for the cell is known. We need the equilibrium temperature T_e to determine α_e from eq.(43). The definition of T_e is as follows; if the system of the atoms and molecules in the cell were suddenly isolated, it would settle down into an equilibrium state with temperature T_e. Now we have

$$0.5(1-\alpha_e)\langle E_2\rangle + \alpha_e\langle E_1\rangle + 0.5\alpha_e D = 0.5(1-\alpha)\bar{E}_2 + \alpha\bar{E}_1 + 0.5\alpha D, \quad (44)$$

where

$$\langle E_2\rangle = \frac{5}{2}kT_e + \frac{k\theta_v}{\exp(\theta_v/T_e)-1},\ \langle E_1\rangle = \frac{3}{2}kT_e,$$

$$\bar{E}_2 = m(\bar{c_2^2} - \bar{c}_2^2) + \bar{\varepsilon}_2^R + \bar{\varepsilon}_2^V,\ \bar{E}_1 = 0.5m(\bar{c_1^2} - \bar{c}_1^2). \quad (45)$$

Here, suffix 2 denotes molecule, 1 denotes atom, and the bars denote average. The energy of the mixture is reckoned from the ground state of molecules at rest. Recall that m is the mass of an atom. The flow velocities $\bar{\mathbf{c}}_1$ and $\bar{\mathbf{c}}_2$ vanish for a homogeneous gas. Equation (43) can be solved with respect to α_e. Substituting α_e into eq.(44), we can find T_e, say, by use of Newton's method. Put the solution T_e into eq.(43), and we have α_e.

6.2 Dissociation as an outcome of T-R-V event

In case of the 2-2 collision, the post-collision quantities can be determined by the procedure in §5. If molecule 1 satisfies the condition (40), then

$$\mathrm{A_2} \longrightarrow \mathrm{A + A}. \quad (46)$$

The velocities $\mathbf{v}_1$, $\mathbf{v}_2$ of the two atoms are determined as follows. The conservation of momentum is

$$\mathbf{v}_1 + \mathbf{v}_2 = 2\mathbf{c}_1' \quad (47)$$

The conservation of energy is

$$mf^2/4 = \varepsilon_1^{R'} + \varepsilon_1^{V'} - D, \quad (48)$$

where $\mathbf{c}_1'$, $\varepsilon_1^{R'}$, $\varepsilon_1^{V'}$ are the post-collision quantities of molecule 1 and $f = |\mathbf{v}_2 - \mathbf{v}_1|$. Once f is obtained from eq.(48), we have

$$\mathbf{v}_2 - \mathbf{v}_1 = f\mathbf{R}^*. \quad (49)$$

The velocities $\mathbf{v}_1, \mathbf{v}_2$ are found from eqs.(47) and (49).

Next we consider the 1-2 collision. Section 5 must be modified since one of the pair is an atom. Before collision, molecule has the velocity $\mathbf{c}_1$, and internal energy ε_1, where $\varepsilon_1 = \varepsilon_1^R + \varepsilon_1^V$, and atom has the velocity $\mathbf{v}_2$. The total collision energy E_c is

$$E_c = E_T + \varepsilon_1 = E_T' + \varepsilon_1', \quad (50)$$

where $E_T = mg^2/3$ and $g = |\mathbf{v}_2 - \mathbf{c}_1|$. The collision temperature T_c satisfies

$$\frac{5}{2}x + \left[\exp\left(\frac{1}{x}\right) - 1\right]^{-1} = \tilde{E}_c,$$

where $x = T_c/\theta_v$ and $\tilde{E}_c = E_c/k\theta_v$. The solution is

$$x = (2/5)\tilde{E}_c \text{ for } \tilde{E}_c \leq 0.25.$$

For other E_c we use Newton's method. The first approximation $x^{(1)}$ is

$$x^{(1)} = (2/5)\tilde{E}_c \text{ for } \tilde{E}_c \leq 2.4,$$
$$x^{(1)} = (2/7)\tilde{E}_c \text{ for } \tilde{E}_c > 2.4.$$

Using the converged T_c, we obtain ζ_v from eq.(33). The probabilities of the elastic, T-R, and T-R-V events are given by eq.(32).

T − R event

The energy E_T' is

$$E_T' = mg'^2/3 = F_c R^{*2/3}, \tag{51}$$

where

$$F_c = E_T + \varepsilon_1^R = E_T' + \varepsilon_1^{R'}.$$

The energy $\varepsilon_1^{R'}$ is equal to $F_c - E_T'$. By use of g' from eq.(51) we have

$$\mathbf{v}_2' - \mathbf{c}_1' = g'\mathbf{R}^*. \tag{52}$$

The conservation of momentum is

$$2\mathbf{c}_1' + \mathbf{v}_2' = 2\mathbf{c}_1 + \mathbf{v}_2. \tag{53}$$

The post-collision velocities $\mathbf{c}_1'$ and $\mathbf{v}_2'$ can be obtained from eqs.(52) and (53).

T − R − V event

Let X be E_T'/E_c. The probability density of X normalized by its maximum is

$$(1+\zeta_v)^{(1+\zeta_v)/2}\zeta_v^{-\zeta_v/2}\sqrt{X}(1-X)^{\zeta_v/2}.$$

Once X is sampled, we have

$$E_T' = mg'^2/3 = XE_c. \tag{54}$$

Since g' is known from eq.(54), we have $\mathbf{c}_1'$ and $\mathbf{v}_2'$ from eqs.(52) and (53). The energy $\varepsilon_1'(= \varepsilon_1^{R'} + \varepsilon_1^{V'})$ is, from eq.(50)

$$\varepsilon_1' = E_c - E_T'.$$

This is divided into $\varepsilon_1^{R'}$ and $\varepsilon_1^{V'}$;

$$\varepsilon_1^{R'} = [2/(2+\zeta_v)]\varepsilon_1',$$
$$\varepsilon_1^{V'} = \varepsilon_1' - \varepsilon_1^{R'}. \tag{55}$$

If the condition (40) is satisfied, the molecule dissociates into two atoms. The procedure is given by eqs.(46)-(49).

6.3 Recombination and immediate deactivation

We follow Yoshizawa's idea.[14] Two atoms recombine and become a transient " super-molecule " A_2^* whose internal energy is greater than D

$$A + A \longrightarrow A_2^*. \tag{56}$$

The super-molecule is then deactivated in collision with a third particle M

$$A_2^* + M \longrightarrow A_2 + M. \tag{57}$$

First we consider the reaction (56). Let $\mathbf{v}_1$, $\mathbf{v}_2$ be the velocities of the two atoms, and $\mathbf{c}_1$ and ε_1 be the velocity and internal energy of A_2^*. The conservation equations take the forms

$$\mathbf{c}_1 = (\mathbf{v}_1 + \mathbf{v}_2)/2, \tag{58}$$

$$\varepsilon_1 = D + m|\mathbf{v}_2 - \mathbf{v}_1|^2/4. \tag{59}$$

Next, we consider the deactivation (57). The collision must be strongly inelastic. It is most probable that only the T-R-V event occurs. We set $P_{TRV} = 1$ for reaction (57). First the species of the particle M must be specified. We see from eq.(36) that the probability that A_2^* collides with some molecule is proportional to $n_2\ \sigma_{22}$ and the probability that it collides with some atom is proportional to $n_1\ \sigma_{21}$. Then the conditional probability that if A_2^* collides, its partner M belongs to species s is

$$\frac{N_s \sigma_{2s}}{N_1 \sigma_{21} + N_2 \sigma_{22}}, \quad (s = 1, 2). \tag{60}$$

Hereafter, $\sigma_{22} = \sigma_{21}$ is assumed for simplicity. Note that A_2^* is not counted in the number N_2. When the species s is sampled according to (60), the collision partner M is chosen at random from N_s particles. If M is molecule, the post-collision state is calculated by the procedure in §5. If M is atom, it is calculated by the procedure of the T-R-V event in §6.2.

6.4 Reaction in spatially homogeneous gas

As an example, let us consider a mixture of atomic and molecular oxygen. The density ρ and equilibrium temperature T_e are chosen as

$$\rho/\rho_D = 10^{-6}, \quad T_e = 5000 \text{ K}.$$

Equation (43) gives $\alpha_e = 0.8847$. The degree of dissociation, α, changes with time until it becomes equal to α_e. The mixture has the energy E_0 given by

$$\frac{E_0}{kN_T} = \frac{1-\alpha_e}{2}\left[\frac{5}{2}T_e + \frac{\theta_v}{\exp(\theta_v/T_e) - 1}\right] + \alpha_e\left(\frac{3}{2}T_e + \frac{\theta_D}{2}\right),$$

where $N_T (= 2N_2 + N_1)$ is the total number of atoms in the mixture. The initial conditions are: there are no atoms; molecules have no vibrational energy; the

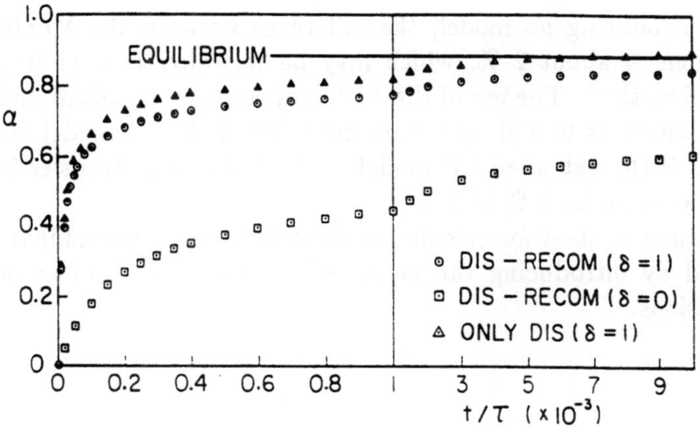

Fig.11 Degree of dissociation. The time scale is changed at $t/\tau = 10^3$.

translational and rotational modes are in equilibrium at T_0, which is given by

$$T_0 = \frac{4}{5}\frac{E_0}{kN_T}(= 27120\ \mathrm{K}).$$

The parameter α in eq.(32) is set to unity. The parameter δ in the dissociation condition (40) is chosen to be 0 or 1. The degree of dissociation is shown in Fig.11, where the result obtained by forbidding recombination is also shown for $\delta = 1$. In case of $\delta = 0$ the approach to equilibrium appears to be too slow. We think that a real δ is near 1. If recombination is forbidden, α exceeds α_e at a certin time. Here time is nondimensionalized by dividing by the mean free time

$$\tau = 2m/\rho\sigma_{22}(6RT_e)^{1/2}$$

of the 2-2 collision at time zero.

7. Conclusions

Accuracy of the variable-hard-sphere (VHS) model, the Larsen-Borgnakke (LB) model, and the extended Larsen-Borgnakke (extended LB) model used in the Monte Carlo direct simulation method (MCDS) is examined by computing hypersonic Couette flows at very small Knudsen numbers and comparing the solutions with the exact solutions of the Navier-Stokes (NS) equation. It is concluded as follows.

(1) The profiles of velocity and temperature obtained by use of the above models generally agree with the NS solutions.

(2) The heat-flux to shear-stress ratio differs from the NS solution at most by 1 %.

(3) The shear stress is sensitive to the employed model. Even if the MCDS

method is used by introducing no model, the difference between the Monte Carlo and NS solutions is about 2 %, which may be regarded as the error inherent in the MCDS method. The use of the VHS model causes no additional error. The difference amounts to 4 % by the introduction of the LB model for T-R energy transfer. If the extended LB model for T-R-V energy transfer is used, the difference increases by 3 % to 7 %.

In addition, a simple method for calculating dissociating and recombinating flows is proposed by introducing the model which gives the number of recombinational collisions.

Acknowledgments

The author wishes to express his thanks to Mr.Takahisa Honda for producing computer graphics from the Monte Carlo data. This research was carried out with the support of the Grant-in-Aid for Scientific Research on Priority Area " Computational Fluid Dynamics " of Japanese Ministry of Education, Science and Culture.

References

[1] Nanbu, K., Proc. 15th Int. Symp. Rarefied Gas Dynamics, (Teubner, 1986), Vol.1, pp.369-383.
[2] Borgnakke, C. and Larsen, P. S., J. Compu. Phys., Vol.18 (1975), pp.405-420.
[3] Bird, G. A., J. Compu. Phys., Vol.25 (1977), pp.353-365.
[4] Harvey, J. K., Thermophysical Aspects of Re-Entry Flows, (eds., J. N. Moss and C. D. Scott, AIAA, 1986), pp.25-43.
[5] Liepmann, H. W. and Roshko, A., Elements of Gasdynamics, (John Wiley, 1960), pp.306-313.
[6] Nanbu, K., Igarashi, S., and Watanabe, Y., Proc. Soviet Union - Japan Symp. Compu. Fluid Dynamics, (USSR Academy of Sciences, 1989).
[7] Ivanov, M. S. and Rogasinsky, S. V., Sov. J. Numer. Anal. Math. Modelling, Vol.3 (1988), pp.453-465.
[8] Koura, K., Phys. Fluids, Vol.29 (1986), pp.3509-3511.
[9] Nanbu, K. and Watanabe, Y., Rep. Inst. High Speed Mech., Tohoku Univ., Vol.43 (1981), pp.21-54.
[10] Anderson, D. A., Tannehill, J. C., and Pletcher, R. H., Computational Fluid Mechanics and Heat Transfer, (McGraw-Hill, 1984), pp.249-250.
[11] Nanbu, K., Honda, T., and Igarashi, S., J. Thermophysics and Heat Transfer (to be published).
[12] Bird, G. A., Proc. Rarefied Gas Dynamics, Pt.1 (AIAA, 1980), pp.239-255.
[13] Nanbu, K., J. Phys. Soc. Jpn., Vol.49 (1980), pp.2050-2054.
[14] Yoshizawa, Y., Rarefied Gas Dynamics, Vol.1, (Academic, 1967), pp.641-660.

Intrusive and Nonintrusive Measurements of Flow Properties in Arc Jets

Carl D. Scott
NASA Johnson Space Center
Houston, Texas 77058

Abstract

Diagnostics of arc jet flows has been a difficult task for over three decades. This paper addresses determining flowproperties of arc jets using both intrusive and nonintrusive techniques. The task of dermining the total enthalpy of the flow continues to be a problem, and several techniques for measuring the enthalpy are discussed. Methods for measuring the heat fluxes are slug calorimeters and Gardon gage instruments. These measurements along with stagnation point pressure measusrements can be used to help determine the enthalpy. Other intrusive techniques for measuring flow properties discussed in the paper are electrostatic probes for determining electron density and temperature, mass spectroscopic measurements of species with gas sampling probes for determining relative species concentrations, and enthalpy probes. Examples of applications of these techniques to arc jets are given. Nonintrusive techniques do not perturb the flow and yield local flow properties. Various spectroscopic techniques, such as atomic and molecular emission, can be used to determine excitation temperature of species and electron density. These techniques can also be used to determine the population of states of species. Examples of various arc jet applications using helium, argon, nitrogen, and hydrogen as diagnostic gases are given. Laser scattering techniques such as laser induced fluorescence and Thomson scattering as applied to arc jets are discussed. Finally, brief mention is given of microwave determination of electron density and the use of photography for flow visualization.

1. Introduction

Testing of spacecraft thermal protection materials and systems to determine their performance is usually done in an arc heated wind tunnel or test facility. Samples of the material are immersed in the high enthalpy flow and heated to some given temperature or at some level of heat flux to simulate the local flight environment. Test articles, constructed to simulate local spacecraft structures, are subjected to heating profiles that test the capability of the materials and structures to withstand the anticipated local flight environment. How well the test environment simulates the flight environment is determined by predicting the flight environment and comparing the predictions with the measurements of the arc jet flow environment. Determining the arc jet environment requires complex analytical and experimental procedures. The accuracy of the simulation often depends on the ability to measure the flow properties and to control these properties within the capability of the facility. This paper addresses intrusive diagnostic techniques such as pressure, heat flux, gas sampling mass spectrometer, and electrostatic probes; as well as nonintrusive diagnostic techniques, such as emission and absorption spectroscopy and laser induced fluorescence (LIF).

The use of arc jets with flow enthalpy that simulates reentry into the earth atmosphere dates back to the 1950's and many of the techniques in use today to determine the flow characteristics date back to that time. In the 1960's spectroscopic and electrostatic techniques were applied to diagnose the flow. However, the use of these sophisticated techniques fell into disuse due to the major commitment of manpower required, a lack of funds, and the high priority of using the facilities to test materials and structures over understanding the details of the flow. The main types of diagnostics or characterization of the flows are heat flux, surface temperature, enthalpy, and pressure measurements. Even the enthalpy was relegated to a subordinate parameter as far as thermal testing of structures was concerned. However, when it became apparent that the Space Shuttle Orbiter thermal protection tiles and the leading edge material were fairly noncatalytic with respect to atom recombination there was increased interest in understanding the composition of the flow and its interaction with thermal protection materials. This has led to the implementation of modern spectroscopic and laser diagnostic techniques.

1.1 Scope

A comprehensive review of arc jet and high temperature flow diagnostics was given by Fauchais in notes for the 1st Joint Europe/US Short Course on Hypersonics, Paris, France, December 7-11, 1987. Much of the present paper covers similar material, but deals with different applications. The two papers are therefore complemetary, and one should refer to Fauchais' notes for completeness.

Arc jet facilities have been applied to various flight regimes - from the high pressure and heating environments of ballistic missile reentry and planetary atmosphere entry to the relatively low pressure environment associated with reentries of lifting manned spacecraft. Most of this paper addresses the lower pressure regime because the author is most familiar with it and because of the current interest in hypersonic manned flight. Although the atmosphere of primary interest is air, the arc jet flow of noble gases will be addressed briefly, mainly because these gases are simple and illustrate the diagnostic techniques and because noble gases may be used as trace species for diagnostic purposes.

In the lower pressure regimes the flow is not in chemical equilibrium and in many instances is not in thermal equilibrium. For example, the electrons may be at a different temperature from the heavy particles. Also, the vibrational state populations may deviate from a Boltzmann distribution. These nonequilibrium effects tend to complicate the flow description and diagnostics. Therefore, to characterize the flow fully it is necessary to measure additional flow parameters. Measurements to determine these nonequilibrium phenomena and effects include spectroscopy, laser induced fluorescence, direct mass spectrometry, and catalytic and electrostatic probes.

2. Measurement Parameters

2.1 Arc Heater Operational Parameters

Basic information about the operation of the arc heater itself is necessary to an understanding of the flow properties. In particular, the total pressure and enthalpy of the flow is derived from measurements of the arc heater operating conditions. (A discussion of the various methods used to determine the average total enthalpy of the flow is given by Pope.[1]) The bulk total enthalpy is obtained from the heat balance method. For the heat balance the arc current, arc voltage, and mass flow rate of the gas are measured. Also, the heat lost to the walls of the heater is determined from the heat gain of the coolant. From these measurements the total enthalpy averaged over the flow is calculated using

$$h_T = \frac{IV - m_c C_p[\Delta T - \Delta T_o]}{m_g} \tag{1}$$

where the current is I, the voltage drop across the heater is V, m_c is the coolant flow rate, C_p is the specific heat of the coolant, ΔT is the temperature rise of the coolant, m_g is the gas flow rate through the heater, and ΔT_o is the temperature increase of the coolant due to viscous effects in the coolant passages. Measurement of the temperature rise of the coolant with no arc yields ΔT_o. Since the temperature rise of the coolant is usually small it is necessary to take some precautions to measure it accurately. This is the major source of uncertainty in determining the bulk average enthalpy. Since Eqn. (1) yields a bulk average it does not reveal any information about the detailed enthalpy distribution in the flow field. To determine the enthalpy of the inviscid core of the jet only heat losses in the heater itself should be included in (1), and not the heat loss to a nozzle wall. To determine the mean enthalpy of the fully developed flow, as in a duct, the energy losses to the duct wall should be included. Also the nozzle losses should be included if one is to use the enthalpy integral (heat flux) method discussed below.

Other techniques for determining the enthalpy of the flow are the sonic throat method (also called the sonic flow method) and the heat flux method. The sonic throat method will be addressed first. In its simplest form the method depends on the assumption of perfect gas flow[1] or equilibrium flow[2]. However, if a nonequilibrium flow code is available the method could be extended to nonequilibrium flow, provided some reasonable starting conditions can be assumed — an assumption not always feasible. The total enthalpy of perfect gas flow is related to the stagnation pressure and the nozzle throat area by the relation

$$\frac{m_g}{p_t A^*} = \left(\frac{1}{h_t} \frac{\gamma^2}{\gamma - 1} \left(\frac{2}{\gamma + 1} \right)^{\frac{\gamma+1}{\gamma-1}} \right)^{1/2} \tag{2}$$

The total pressure in an arc heater may be used to determine the enthalpy or total temperature if the flow can be considered in equilibrium upstream of the nozzle throat. Accurately determining the total pressure also needs careful attention. At the upstream end of the heater where the gas velocity of the flow is near zero the stagnation pressure may be measured by a static pressure tap. However, this stagnation pressure is not the same as the stagnation pressure of the flow at the nozzle throat due to the heating and viscous effects in the arc heater.[3] A static pressure measurement at the nozzle throat is not an accurate measure of the stagnation pressure, since the flow is near sonic velocity. Relating the static pressure to stagnation pressure requires knowledge of the effective ratio of specific heats and the Mach number. The Mach number at the nozzle throat is usually taken to be unity, but the location of the sonic line may not be accurately known due to the viscous boundary layer affecting the effective throat area. Another *caveat* to be aware of is that there is often some slag or other contaminants in arc heaters which may foul or clog the orifice of the pressure tap and interfere with accurate measurements of the pressure. These considerations often make determining the enthalpy by the *sonic flow* method difficult.

Another difficulty with bulk methods such as the sonic flow and the energy balance enthalpy methods is that the flow in arc heaters is usually not very uniform. Therefore, a local measure of the enthalpy is needed. One can make a determination of the local enthalpy using the heat flux method.[1,4] Since the heat flux to an object depends on the total enthalpy of the flow one can solve a heat flux relation for the enthalpy. In the past, the Fay and Riddell[5] heating equation or a similar relation has been used to infer the enthalpy. In using this technique the catalytic recombination efficiency of the surface of the heat flux sensor must be known or should be very large (effectively fully catalytic). The enthalpy then can be obtained from the measured heat flux and stagnation point pressure

$$h_T = C\, q_s \left(\frac{1}{R_{eff}\, p_s}\right)^{1/2} \tag{3}$$

where q_s is the stagnation point heat flux, C is a constant, R_{eff} is the effective radius of the heat flux measurement device, and p_s is the stagnation point pressure measured at the same location as the heat flux. The effective radius of the heat flux probe depends on the shape. Most probes are either spheres or flat face cylinders. For spheres, the effective radius is simply the sphere radius. For flat face cylinders the effective radius must be determined analytically or by experiment. Hiester and Clark[4] found for the conditions of their series of tests in a number of arc jet facilities that

$$R_{eff} = R_{cyl}/0.54 \tag{4}$$

where R_{cyl} is the radius of the cylinder. Actually, the factor, which depends on the velocity gradient, is a function of density ratio across the shock wave. Bade and Yos,[6] in their nonequilibrium arc jet flow analyses, use a correlation of the measurements of Boison and Curtiss[7] for a flat face cylinder.

Pope[1] compares the heat flux method with bulk flow methods by surveying the stream and integrating the enthalpy distribution across the stream using the relation

$$\frac{h_o}{\bar{h}} = \frac{m_g}{\pi \int_0^R \frac{h}{h_o} \rho u r dr} \tag{5}$$

where h_o is the centerline enthalpy, $\bar{h}$ is the bulk average enthalpy, h is the local enthalpy, ρu is the local mass flux (assumed to be constant function of radius), and the integration is over the area of the flow cross section.

There has been much uncertainty in the enthalpy determined from the various methods as can be seen in Figs. 1 and 2 from Pope's article[1] and Fig. 3 from Hiester and Clark[4]. Therefore, better techniques would be helpful. Nonintrusive techniques will give some of the parameters that make up the enthalpy in a nonequilibrium flow, such as the velocity and temperatures, but they are not likely to yield all modes of energy, in the near future. Intrusive techniques, such as electrostatic probes and mass spectroscopic probes yield additional information, such as, the electron temperature and species concentrations respectively. However, these intrusive techniques may perturb the flow and have other uncertainties associated with them.

Fig. 1

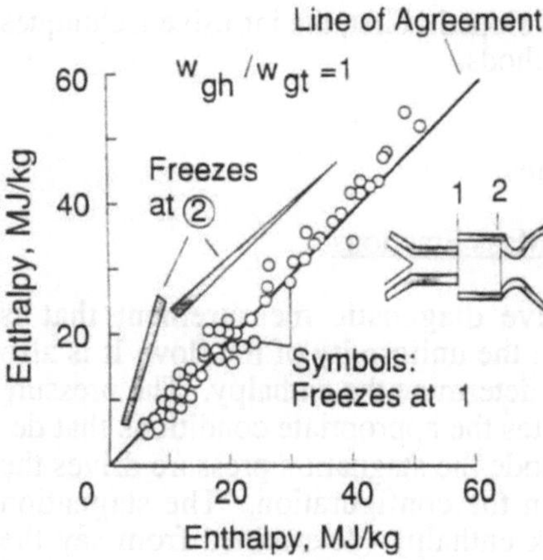

Fig. 2

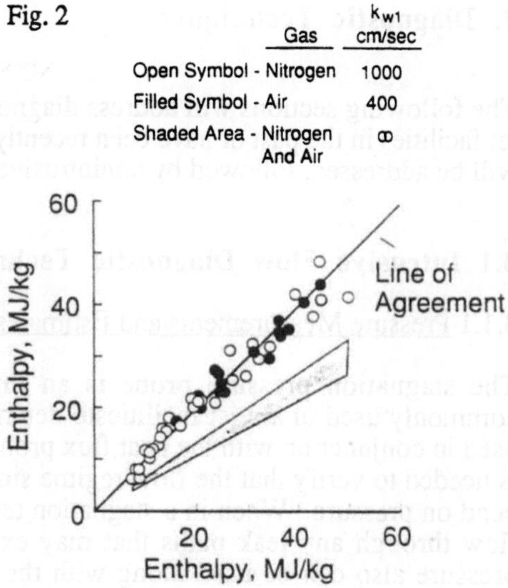

Fig. 1 Comparison of enthalpy measurements — sonic flow and stream enthalpy profile methods. (From Pope[1])

Fig. 2 Comparison of enthalpy measurements — sonic flow and heating-rate methods. (From Pope[1])

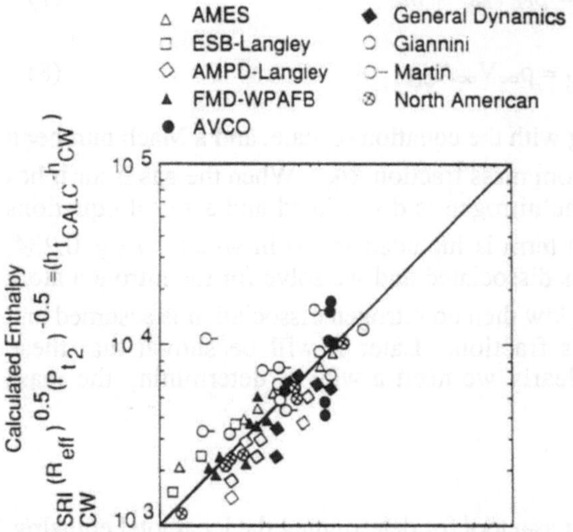

Fig. 3 Enthalpy calculated from heating rate versus sonic flow enthalpy. (From Hiester and Clark[4])

3. Diagnostic Techniques

The following sections will address diagnostic techniques that have been used in arc jet facilities in the past or have been recently developed. First, the intrusive techniques will be addressed, followed by nonintrusive methods.

3.1 Intrusive Flow Diagnostic Techniques

3.1.1 Pressure Measurements and Estimates of Mass Fractions

The stagnation pressure probe is an intrusive diagnostic measurement that is commonly used in arc jet facilitiesto determine the uniformity of the flow. It is also used in conjunction with the heat flux probe to determine the enthalpy. The pressure is needed to verify that the flow regime simulates the appropriate conditions that depend on pressure. When in a stagnation test mode the stagnation pressure drives the flow through any leak paths that may exist in the configuration. The stagnation pressure also can be used along with the bulk enthalpy (determined from say the energy balance method) and the gas flow rate to estimate the amount of dissociation.[8] This technique assumes that the flow is uniform and that the effective area of the nozzle or inviscid core is known. When a single gas such as nitrogen is used the dissociation is found by a straightforward use of the conservation equations

$$h_T = \frac{1}{2} V_\infty{}^2 + \alpha_N h_N + C_p T_\infty \qquad (6)$$

$$p_{T2} = \rho_\infty V_\infty{}^2 + p_\infty \qquad (7)$$

$$m_g = \rho_\infty V_\infty A_{eff} \qquad (8)$$

These equations can be solved along with the equation of state, and a Mach number a Mach number relation, to find the atom mass fraction α_N. When the gas is air it has been assumed that either none of the nitrogen is dissociated and a set of equations analogous to (6)-(8) is solved; or a term is included in (6) in which $\alpha_O = 0.234$, which is to say that all the oxygen is dissociated and we solve for the nitrogen atom mass fraction α_N. If the enthalpy is low then no nitrogen dissociation is assumed and we solve for the the oxygen mass fraction. Later it will be shown that these assumptions may not be valid. Clearly we need a way of determining the mass fractions directly.

3.1.2 Heat Flux Measurements

Heat flux probes have been discussed as a tool for determining the local total enthalpy. The heat flux is also used to set up test conditions for materials and structures testing. Usually the heat flux sensors are installed in models that have the same geometry as the test article. They are located at strategic locations at which the incident heat flux is to be imposed on the test article — frequently the stagnation point.

3.1.2.1 Heat Flux Sensors

Heat fluxes are commonly measured by two types of instruments in arc jet facilities where the heat flux ranges from about 1 W/cm^2 to more than 500 W/cm^2. For the

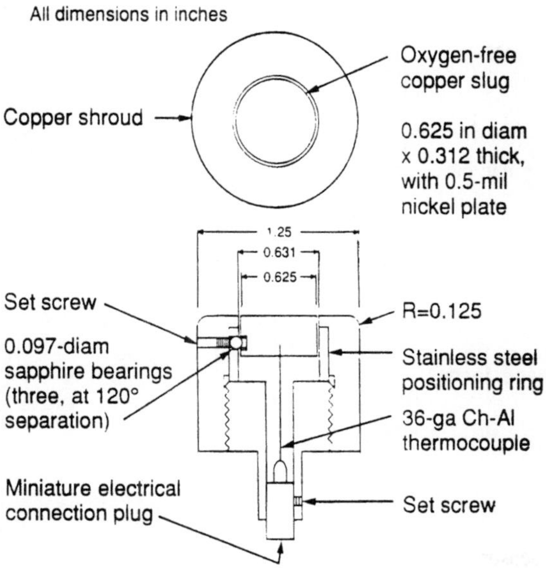

Fig. 4 Typical slug heat flux sensor. (From Hiester and Clark[4])

most accurate work the slug calorimeter is often used because its output does not depend on a calibration. The response of the sensor is strictly a function of the geometry and physical properties of the slug, provided the slug is properly isolated from the supporting hardware. The heat flux q is determined from the temperature rise rate of the slug of material, usually copper.

$$q = mC_p\frac{dT}{dt} \qquad (9)$$

where m is the mass of the slug, and dT/dt is the rise rate of the slug temperature. An example of a slug type sensor is shown in Fig. 4 which is taken from Ref. 4. This type of sensor has been successfully used in many applications in which the surface coating could be varied to assess catalytic properties at low temperatures.[9] Also, slug calorimeters were used to measure the heat flux to nickel and Teflon surfaces to assess slip and catalytic boundary equations[10] for a time dependent relaxation computational fluid dynamics code[11].

Another type of instrument for measuring heat flux in arc jet facilities is the Gardon gage shown schematically in Fig. 5. The principle on which heat flux is determined is that the temperature difference between the center and outer edge of the constantan disk is proportional to the heat flux. This temperature difference is sensed by a pair of differential thermocouples, one at the junction between the copper wire and the center of the disk, and the other at the junction between the outer edge of the disk and the copper body of the instrument. The output then can be calibrated with a suitable standard heat flux source, such as a black body. The sensitivity of the instrument is determined by the thickness of the constantan disk and its radius. These instruments are commercially available and come in various ranges. They may be water cooled for continuous operation. One drawback to the Gardon gage is its sensitivity to electrical noise generated by the arc. Therefore, care must be taken to electrically isolate its leads by shielding and to avoid ground loops. Another drawback to gages that require calibration is that the calibration source is usually a radiant source, thus the radiant

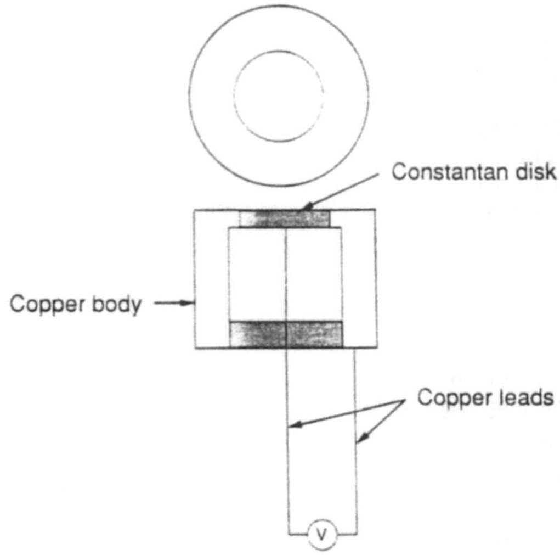

Fig. 5 Schematic of Gardon gage heat flux sensor.

absorbance of the gage must be known. Usually this is accomplished by coating the gage with carbon black for the calibration. However, when using the probe in a dissociated nonequilibrium flow the catalytic recombination rate of the surface must be known or it must be fully catalytic. This means that the probe surface must be carefully cleaned and have a coating of a material of known catalycity. Frequently one assumes that the gage is fully catalytic, but that is only an approximation. Slug calorimeters avoid the problem of having to put on an overcoat of carbon black for calibration. However, since they are transient devices they cannot be used continuously or for many insertions into the flow without allowing to cool down.

3.1.3 Enthalpy Probes

In the early 1960s enthalpy probes[12] were proposed for determining the local stream enthalpy. An enthalpy probe continually draws a sample of the stagnation flow into the probe and measures its heat loss within the probe. However, it was found that these probes were difficult to use because the heat loss was so small compared with the heat transferred to the probe at its tip from the external arc jet flow. The amount of coolant required to cool the probe was so great that the temperature rise of the coolant was extremely small, and hence was sensitive to small errors. Enthalpy probes did not work well in low pressure environments associated with manned vehicle thermal protection material testing. Examples of enthalpy probe designs are shown in Fig. 6.

3.1.4 Electrostatic Probes

Electrostatic probes are conductors in contact with the flow to which a positive or negative bias voltage is applied. The current collected due to flux of electrons or ions is measured as a function of the applied voltage. If the probe circuit is left open it will acquire a potential, called the floating potential, that results in a net zero current. The flux of ions and the flux of electrons balance each other and the net flux is zero. To acquire this floating potential the probe will have a net negative charge on it such that the electrons, which are much more mobile than the ions, will be repelled and the ions will be attracted in such a way that the net fluxes are equal. As the probe bias is made more negative it will collect more ions than electrons. At large negative bias the probe

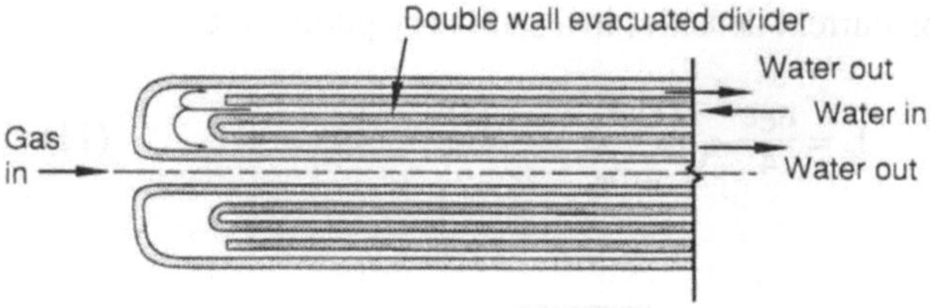

AVCO Corp - Split flow with insulated divider
(tare measurement required)

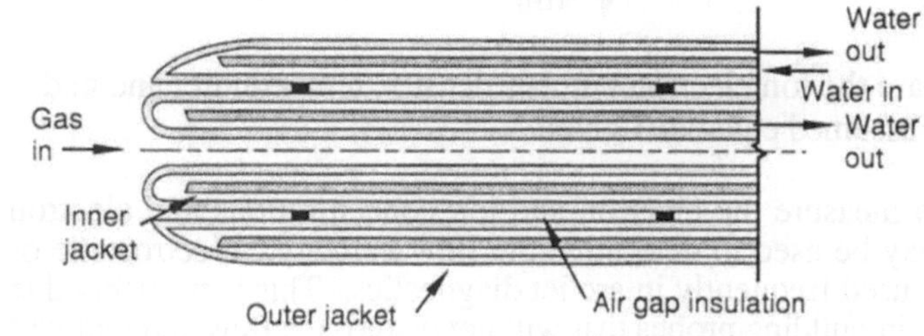

Greyrad Corp - Model G-13, calormeteric flow separated from
outer coolant flow (tare measurement not required but advisable)

Fig. 6 Typical enthalpy probe designs.

will reach what is called ion saturation. That is, no more ions can reach the probe because they cannot diffuse fast enough into the "sheath" surrounding the probe. The sheath is a region in which there may be charge separation and the plasma is no longer neutral. The sheath thickness depends on the applied potential; and in arc jet conditions, where the electron density is relatively high ($> 10^{10}$ cm^{-3}), the sheath is often very thin with respect to other dimensions in the flow, including the mean free path. In the thin sheath case we say that the sheath is collisionless, making the analysis much simpler. As the probe bias is made more positive the probe will collect more electrons, and at sufficiently high potential will reach its electron saturation current. In analogy with the ions, the electron current saturates when the diffusion of electrons into the sheath no longer can keep up with their collection by the probe. Since the electrons are so light, and hence have a much higher thermal velocity compared with the ions, the electron saturation current is orders of magnitude larger than the ion saturation current.

In the transition zone between ion saturation and electron saturation the net current depends strongly on the bias voltage and on the electron temperature. The slope of the logarithm of the current as a function of bias voltage is proportional to the inverse of the temperature, that is

$$\log(J_+ + J_-) = \log J_- - \frac{1}{T_e}\frac{eV}{k} \qquad (10)$$

where $J_{+,-}$ are the ion and electron current densities, T_e is the electron temperature, and V is the bias voltage. We see that the temperature can be determined from the slope. The reason the slope of this equation decreases with increasing electron temperature is because more of the electrons have velocity (energy) great enough to overcome the negative bias of the probe and therefore the probe is able to collect more electrons (J_-) at large negative bias.

The ion and electron saturation current densities in a stationary plasma are

$$J_- = \frac{n_e e}{4}\sqrt{\frac{8kT_e}{\pi m_e}} \tag{11}$$

and

$$J_+ = 0.4 n_i e \sqrt{\frac{2kT_e}{\pi m_i}} \tag{12}$$

where $n_{i,e}$, $T_{i,e}$, $m_{i,e}$ are the ion/electron number density, temperature, and and mass, respectively. T_i is assumed equal to T_e here.

Electrostatic probes also measure the electron and ion concentrations and electron temperature; and they may be used to determine the flow velocity. Electrostatic or Langmuir probes are not used frequently in arc jet diagnostics. This is probably due to the degree of difficulty in building probes that will not disturb the flow and that will survive the high heating and oxidizing environment. Also the theory of the probes often requires information about the flow, e. g. the density, velocity, etc that may not be known or must be determined by other means. However, this technique may reveal a lot of information about the spatial distribution of ions in the flow and the electron temperature that can be used with spectroscopic information to assess the rate of formation of species excited by electrons. If the flow is static or the flow velocity is much less than the random thermal velocity of the ions and electrons then static probe theory can be used to analyze the current to the probe.

Water cooled conical probes and uncooled cylindrical probes found use at the Langley Research Center.[13] (See Fig. 7, where a conical probe design is shown.) Electrostatic probes have been used in arc generated argon plasma jets[14,15] and also in air[13] and nitrogen.[13,16,17]

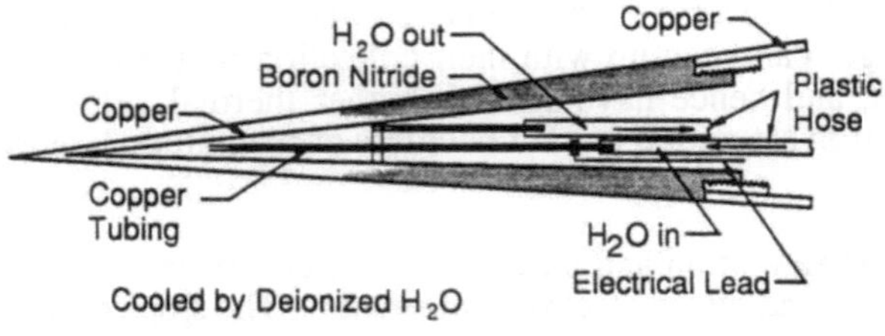

Fig. 7 Conical water cooled electrostatic probe for arc jet service. (From Duckett and Jones[13])

3.1.5 Gas Sampling Mass Spectrometer Measurements

The final intrusive diagnostic technique that will be discussed is the use of mass spectrometer analysis of samples of flow drawn in from a stagnation point gas sampling probe. The first use in arc jets of such a probe, to the author's knowledge, was made about the same time in different facilities by Willey[18,19] and Lasgorceix, et al.[17] Willey applied the method to measuring the relative concentrations of the constituents of air in a supersonic arc jet flow by sampling with a water cooled quartz-tipped probe and a water cooled copper probe adapted from a Pitot pressure probe. Fig. 8 is a schematic of Willey's apparatus. He showed that it is necessary to make

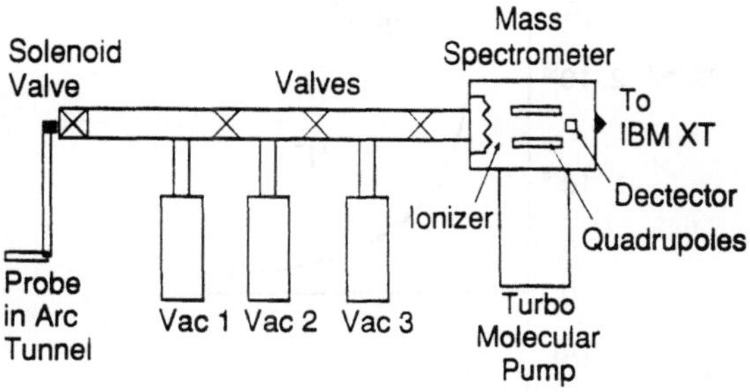

Fig. 8 Schematic of gas sampling probe used with mass spectrometer for species analysis. (From Willey[19])

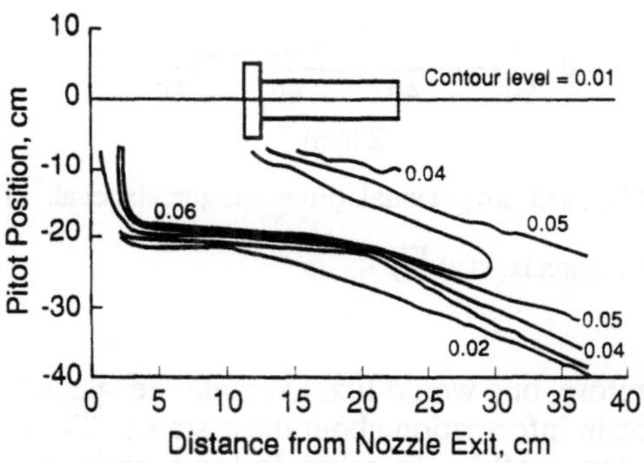

Fig. 9 Nitrogen atom contours in arc jet flow around a blunt body. (From Willey[19])

the probe from noncatalytic materials such as quartz, and to make the length of the tube to the mass spectrometer as short as possible. He demonstrated that the concentrations of atomic species in the flow varies with axially and radially, in contradiction with one dimensional nonequilibrium flow calculations[6]. See Fig. 9, which shows the contours of nitrogen atoms in the vicinity of a blunt body inserted into the flow. Lasgorceix, et al. also showed that the flow in their subsonic nitrogen jet was significantly out of equilibrium and was not frozen. See Figs. 10 and 11 which show the decrease of ion and neutral concentrations with axial distance which contradicts the nonequilibrium predictions. These are very important results and indicate that mass spectrometry should be used to verify the composition of the flow in arc jets when catalysis measurements are made. It also means that the gas reaction rate models need significant refinement.

3.2 Nonintrusive Diagnostic Techniques

So far, intrusive techniques that perturb the flow field have been discussed. It is also of great importance to the understanding of the flow that nonintrusive techniques also be used. Intrusive and nonintrusive techniques are in many ways complementary. The nonintrusive techniques to be discussed are mainly spectroscopic. Since arc jets flows are at very high enthalpies, they tend to radiate spontaneously, especially in the shock layers. There is a lot that can be learned from the emission. However, the free stream flow radiates to a lesser extent and not all excited states of species in the flow have spontaneous radiative transitions or have transitions in convenient wavelength ranges. Also there are limitations due to either detector response or due to overlapping

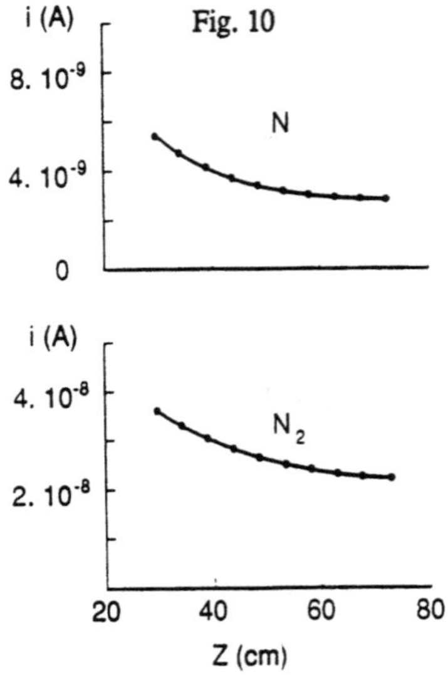

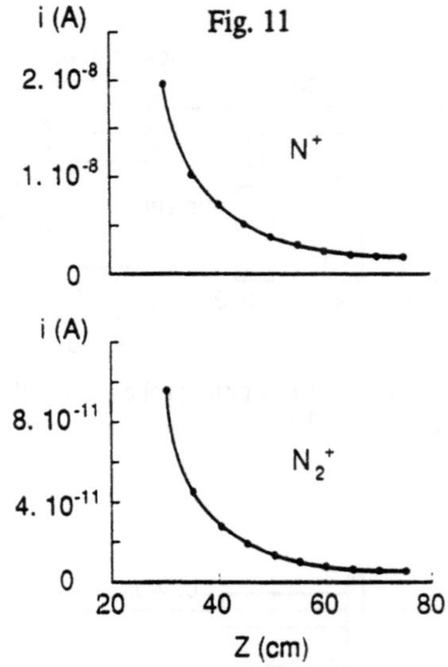

Fig. 10 Axial profile of the neutral particles, N and N_2, and arc potential. (From Lasgorceix, et al.[17])

Fig. 11 Ion concentrations on the jet axis. (From Lasgorceix, et al.[17])

radiation from other species or states. Therefore, one would like to excite the species to radiate from specific states in order to obtain information about these states. Thus, state-selective laser induced transitions are useful. In what follows emission spectroscopy, and to some extent absorption spectroscopy will be discussed. Then we will discuss recent developments in laser induced fluorescence applied to arc jet flows. Briefly we will discuss laser Thomson scattering that has been used to determine electron density and temperature in a plasma jet. We will address briefly microwave propagation experiments that were used to measure electron concentration in an arc jet flow. Finally simple photography will be discussed as useful tool to visualize the flow in self luminous arc jet flows.

3.2.1 Spectroscopic Techniques

Two areas of spectroscopic techniques will be addressed here. The first is a brief description of apparatus and techniques for measuring the spectral radiation. This description addresses some of older methods as well as modern techniques, including calibration sources. The other area of spectroscopic techniques is the interpretation of measurements with suitable theoretical methods. Interpretation of measurements will be the main area of emphasis in this paper since it relates to both emission and absorption spectroscopy as well as to laser induced fluorescence spectroscopy. Examples of techniques applied to arc jets will be given.

3.2.1.1 Spectroscopic Measurement Apparatus

Various emission and absorption spectral instruments have been used in arc jet facilities. These range from simple scanning monochromators to high resolution grating spectrographs and spectrometers. The low resolution monochromators have good light gathering ability because of their short focal length and low F-number (ratio of focal length to aperture diameter). For high resolution it is necessary to sacrifice some light gathering capability because the focal length of the devices must be long. In the 1960s one commonly used[20,21] high resolution spectrograph was a 3.4-meter

Ebert type. Its F-number is about 22, but it can achieve a first order dispersion of about 0.55 nm/mm with a grating ruled with 590 lines/mm. It covers a wide spectral range in one exposure. Today's better grating technology now could improve the dispersion significantly.

The detectors used with spectrometers and monochromators are usually photomultiplier tubes that are sensitive to the wavelengths of interest. Their time response is very fast and they have the advantage that their output can be recorded electronically for subsequent computer manipulation. In some spectrometers more than one photomultiplier can be used at a time to measure the light from several spectral lines. This requires using one exit slit and photomultiplier at each line.

To scan lines to determine their line shape or to obtain a sequence of lines or the variation of continuum with wavelength, the grating may be rotated and a single detector can measure the spectrum as a function of wavelength, provided the source is not changing in time. If the region of the spectrum is large and high resolution is required, this procedure may take a long time. Another method of obtaining a large wavelength range with high resolution is to use a spectrograph with a photographic plate as the detector. One can achieve very high resolution with this technique because with small grain emulsions the resolution is limited only by the entrance slit, whereas, the photomultiplier tube also requires an exit slit which degrades the resolution somewhat. Moreover, the spectrographic plates may be exposed for very long periods of time to achieve sufficient exposure to detect low intensity spectra. There are several difficulties with using photographic detection, the first is that the response is nonlinear and requires a densitometer to read the transmission of the plates to determine intensity. Development of the plate must be highly controlled and the density of the plate must be carefully calibrated for quantitative work. Interpreting the results is very tedious. The sensitivity of photographic emulsions in the IR is fairly low, requiring long exposures. However, the photographic recording of spectra is very good for yielding a comprehensive graphical picture of the spectral characteristics of the flow spectrum over a wide wavelength range.

Modern innovations for measuring spectra include using linear diode array (LDA) detectors. These devices have about 1000 detector elements covering a distance of about 25 mm. These LDAs can acquire 1000 pieces of spectral information at one time. They are then read out sequentially, digitized, and recorded by a multichannel optical analyzer (OMA) for computer data handling and analysis. Their advantage is that they can acquire many channels of data simultaneously and repeatedly; therefore, if the intensity of the source is changing, all wavelengths will be affected to the same extent and the relative intensity of the lines is not changed. The time dependence of the entire spectrum can also be obtained by gating the LDA. One disadvantage is that their resolution capability is somewhat limited; but that can be overcome by selecting gratings with the appropriate blaze and ruling. One is then limited to the wavelength range that can be detected in a single "exposure."

3.2.1.2 Spectroscopic Measurement Calibrations

For quantitative work spectrometers must be calibrated in both wavelength and intensity sensitivity. Wavelength calibration sources are usually gas discharge lamps filled with a gas that radiates with a well known spectrum. These lamps may be excited with a high voltage source via electrodes in contact with the gas, or they are sometimes excited with an radio frequency (RF) electrodeless source. The electrodeless discharge has the advantage of operation at a lower temperature and pressure, therefore, their spectral lines are usually narrower and they are less subject to contamination. The lamps with electrodes are simpler to use, may have higher intensities, and come in a variety of gases to provide a wide coverage of wavelength ranges. Another rather cumbersome source used for very fine wavelength definition is the iron arc. The iron arc spectrum contains thousands of closely spaced lines that have been well catalogued.[22]

Common sources for intensity calibration are the standard tungsten ribbon filament lamps.[23] These can be obtained with calibrations traceable to the National Bureau of Standards. These lamps are good from the near UV to the near IR. For the vacuum ultraviolet one might use a high pressure arc discharge or a carbon arc.[24,25] For calibrating in the IR, globar sources and black body cavities[26] are frequently used.

3.2.1.3 Determining Properties from Spectral Analyses

To determine information about the arc jet flow from the emitted or absorbed spectra one must have a theory that relates the measurements to some property of interest. Spectroscopic techniques actually measure the rate at which the gas emits light, that is the intensity. The intensity depends on two basic quantities, the number of species per unit volume in the particular upper excited state and the probability that the species will make a transition to a lower state. The transition probability A_{ij} is primarily a property of the emitting species and is determined from quantum mechanical calculations or is measured. However, A_{ij} may depend slightly on the environment in which the species exists, that is, collisions and electric and magnetic fields may perturb the electronic energy states slightly and cause some small deviation in the energy levels of the states and their associated transition probabilities. These are small effects, and except when we are interested in line broadening, they usually can be neglected. Suffice it to say that transition probabilities (and oscillator strengths) are properties of the radiating species.

The radiation also depends on the number of species in a particular state and the transition probability. The radiation emitted from a species in state "i" to state "j" is

$$\mathcal{E}_{ij} = N_i A_{ij} h\nu_{ij} \tag{13}$$

where $\mathcal{E}_{ij}$ power radiated per unit volume, h is Planck's constant, and ν_{ij} is the frequency of the photon emitted in the transition from the ith state to the jth state. Equation (13) holds in general — equilibrium or nonequilibrium. When local thermodynamic equilibrium exists the relative number of species in a particular excited state is given by the Boltzmann relation

$$\frac{N_i}{N_o} = \exp\left(\frac{-E_i}{kT}\right) \tag{14}$$

where N_o is the number of species in the ground state, N_i is the number of species in the state "i", E_i is the energy of the state "i" relative to the ground state, k is the Boltzmann constant, and T is the temperature.
Combining (13) and (14), the emitted power corresponding to a transition in an equilibrium gas is then

$$\mathcal{E}_{ij} = N_o A_{ij} h\nu_{ij} \exp\left(\frac{-E_i}{kT}\right) \tag{15}$$

From the relative intensities of atomic lines, continuum, or within molecular bands it is possible to deterimine the excitation temperatures. From absolute intensities it is possible to obtain densities of the excited states and in equilibrium, the total density of the species. These techniques will be discussed in the following paragraphs.

Atomic Line Diagnostics - The temperature or population of states can be obtained from intensities of atomic lines. Equation (15) can be written in terms of the oscillator strength f, which is proportional to the transition probability A_{ij} The intensity I is given by

$$I = \frac{2\pi hc^2 r_o}{\lambda^3} \, g \, f \, N_o \exp\left(\frac{-E_u}{kT}\right) \tag{16}$$

where g is the statistical weight of the upper state.

Equations (15 and 16) reveal that when the states are in equilibrium a plot of the logarithm of the measured $\mathcal{E}_{ij}$ or I versus E_i/k will yield the excitation temperature. The slope of the plot is 1/T. If the population of states deviate from equilibrium then the points will not fall on a straight line.

Argon Lines - Shipley[21] has used this intensity-slope method to determine temperatures in the shock layer of a blunt body in an argon plasma jet flow. A Boltzmann plot of his measurements are shown in Fig. 12. Bowen[27] used the Balmer series of hydrogen to obtain a temperature distribution in a high pressure under expanded jet. Adcock[28] measured the temperature on the centerline of the free stream of an argon plasma jet. He also used a simplification often made when equilibrium is assured. Only two lines (states) are measured and the temperature is found from the ratio of intensities of the two lines.

Copper Lines - This approach has also been used for determining the excitation temperature in atmospheric arc jets using copper lines.[29] Copper from electrode erosion is a contaminant often found in arc jet flows.

Nitrogen Lines - In air and nitrogen jets nitrogen lines are sometimes used to determine the excitation temperature. Park[30] measured the electronic excitation

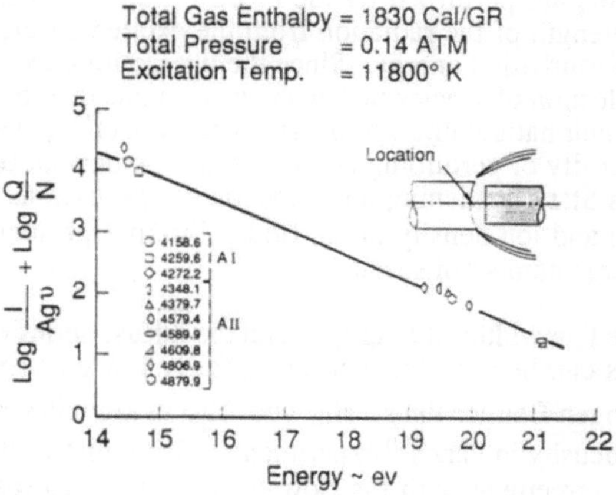

Fig. 12 A Boltzmann plot of argon spectral lines. Both Ar I and Ar II lines are presented. (From Shipley[21])

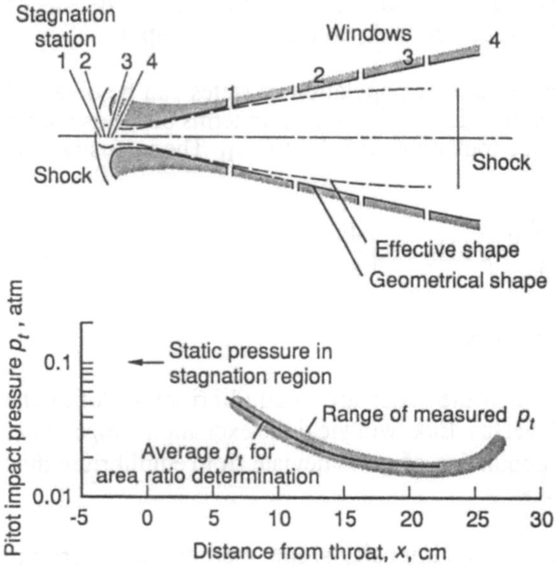

Fig. 13 Park's nozzle shape and impact pressure. (From Park[30])

temperature in a supersonic nozzle expansion that started from a stagnating inlet flow in a nitrogen-hydrogen mixture. The source of the flow was an arc heater. Park's nozzle was inserted in the arc jet flow (Fig. 13.) The measured temperature increased as the flow approached the converging part of the nozzle, then decreased as the flow expanded supersonically. He got good agreement with flow theory calculations.

Helium Lines - Scott[20] determined the temperature profile of a helium arc jet flow using the line intensity technique. His results, using absorption oscillator strengths from Green, et al.[31] and energy levels of Martin,[32] are shown in Fig. 14. From these measurements it was shown that the electrons were in equilibrium with the electronic states of helium.

<u>Stark Broadening</u> - When atoms or molecules undergo collisions with neighboring ions and electrons their energy levels are perturbed by the electric fields of their neighbors. This results in the wavelength of the radiation from the excited species being shifted by the amount of the perturbation energy. Since the interactions are at various distances or strengths the collection of species radiate at wavelengths that may cover a fairly wide range relative to their natural line width. The result is a broadened line, whose width depends on the density of perturbing neighbors, and somewhat on the temperature of the plasma. This Stark boadening phenomenon can be used as a diagnostic to determine the electron and ion density in the flow. Griem[33] presents tables of line broadening data for a large number of species.

H and He Line Broadening - Because line widths of hydrogen and the diffuse series of helium are very broad these species can be used for measuring fairly low electron densities. Line broadening of hydrogen Balmer lines, $H\alpha$ and $H\beta$ in arc jets was used to determine the electron (ion) density in Park's[30] experiments. Using line width data from Griem,[33] Park got good agreement with his flow theory[34] in the throat section of the nozzle and reasonable agreement in the expansion section. His results are shown in Fig. 15.

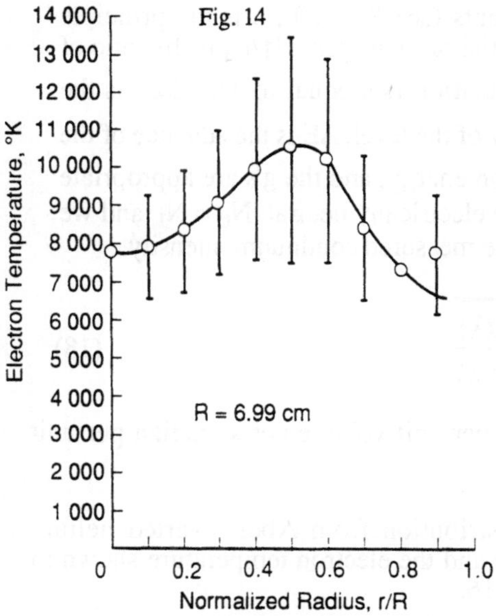

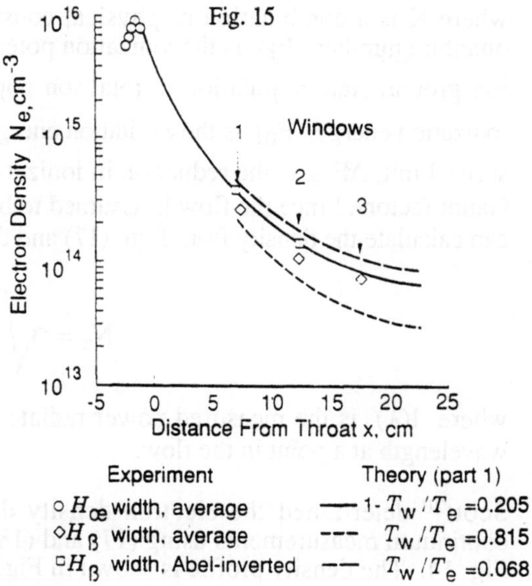

Fig. 14 Electron temperature distribution in helium arc jet flow, determined form Boltzmann plots. Error bars denote standard deviation. (From Scott[20])

Fig. 15 Variation of electron density in nozzle. (From Park[30])

Scott[20] developed a quasistatic theory of line broadening of helium triplet diffuse lines (2^3P - n^3D transitions) and compared these results with measurements in a pure helium arc jet expansion flow. The electron density used in the calculations was determined from continuum measurements to be discussed in a subsequent section.

N and A line Broadening - At higher electron densities nitrogen and argon line broadening may be useful. In general, however, one must use fairly high resolution instrumentation to obtain electron densities using line broadening in low density arc jets. Nitrogen and argon must normally be used with atmospheric pressure plasmas and not with low density arc jets.

<u>Continuum Radiation</u> - The continuum radiation arises from transitions from free electrons to either bound states or to free states. This is called Bremstrahlung radiation. The intensity distribution of the continuum radiation depends on the the electron temperature. The level of radiation depends on the electron density. Scott[20] determined the electron density in a helium arc jet using the relation[35] for the continuum emission coefficient

$$\varepsilon(\lambda,T) = \frac{K}{\lambda^2}\left\{\frac{N_1}{N_i}\sum \frac{g_{nl}(\lambda)}{n^3}\exp\left(\frac{E_\infty - E_{nl}}{kT}\right)\right.$$

$$\left. + \sum \frac{g_n(\lambda)}{n^3}\exp\left(\frac{E_H}{n^2 kT}\right) + \frac{g_f(\lambda,T)kT}{2E_H}\exp\left(\frac{\Delta E}{kT}\right)\right\}$$

$$\times\ N_e N_i \exp\left(-\frac{E_\lambda + \Delta E_\infty}{kT}\right)\left(\frac{E_H}{kT}\right)^{\frac{3}{2}} \qquad (17)$$

where K is a combination of physical constants (see Ref. 20), n is the principle quantum number, E_H is the ionization potential of hydrogen, N_1/N_i is the ratio of ion ground state population to total ion population (set equal to 1), E_∞ is the ionization energy, E_{nl} is the excitation energy of the level, ΔE is the advance of the series limit, ΔE_∞ is the reduction in ionization energy, and the g's are appropriate Gaunt factors. Since the flow is assumed to be electrically neutral $N_e = N_i$ and we can calculate the density from Eqn. (17) and the measured continuum intensity.

$$N_e = \sqrt{\frac{I(\lambda)}{\varepsilon(\lambda)}} \tag{18}$$

where $I(\lambda)$ is the measured power radiated per unit volume per steradian per unit wavelength at a point in the flow.

Scott[20] determined the electron density distribution from Abel inverted helium continuum measurements using (17) and (18) and the electron temperature shown in Fig. 14. The density profile is shown in Fig. 16.

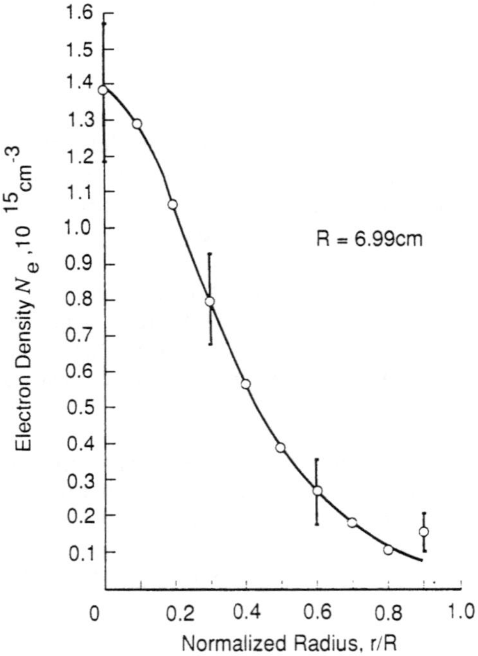

Fig. 16 Electron density distribution in helium arc jet flow. Error bars denote standard deviation. (From Scott[20])

3.2.1.3.4 Molecular Radiation

Because of the complex configuration of electrons in molecules and the additional modes of energy, *viz.* the rotational and vibrational modes, molecular radiation is much more complex than atomic radiation. See Fig. 17 for an example of the visible

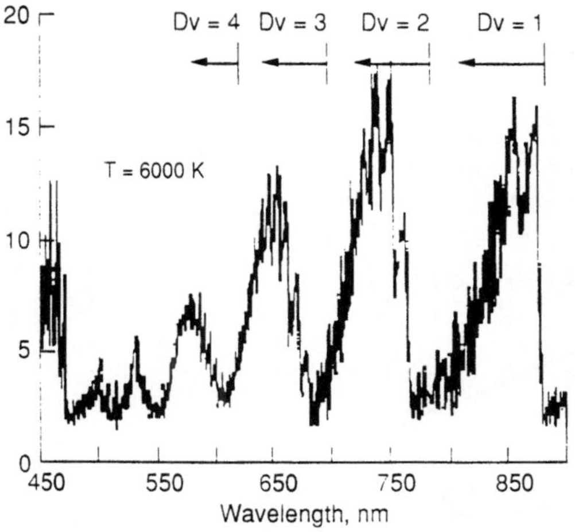

Fig. 17 Computed spectrum of N_2^+ for $T = T_v = T_e = 6000$ K. (From Blackwell, et al.[41])

radiation spectrum of a simple diatomic molecule, nitrogen. One can see that there are many complex band features that make up the spectrum. The rotational structure of these band can be seen in highly resolved spectra; and from the pattern of intensities the rotational temperature can be obtained. We will not go into the details of the theory of the spectra except to say that the intensities are determined from equation (15). The complexity arises from the form of the transition probability A_{ij} and the form of the energy E_i. The states i and j depend on many quantum numbers[36], e. g. the electron spin, the orbital angular momentum, total angular momentum of the electrons, component of angular momentum along the internuclear axis, the rotational quantum number, and the vibrational quantum number. To use the radiation from even the simple molecule such as N_2^+, features of a complex spectrum must be isolated and studied to address one mode of energy, while keeping the others fixed. That is, if we address the intensity variation of the spectrum as a function of the rotational quantum number we can obtain the rotational temperature. As before, a temperature is defined only if the energy levels are connected by a Boltzmann factor distribution. For rotation, which requires only 5 to 10 collisions to come to equilibrium, equilibrium usually exists in conditions of arc jet flows. On the other hand, vibration relaxes more slowly, requiring many more collisions. Therefore, vibrational states may be out of equilibrium, especially in the expanding low density flow of arc jets and near shock waves, although the situation may be somewhat nearer vibrational equilibrium in the shock layer of a blunt body.

Molecules of interest in arc jets used for reentry simulation are N_2, N_2^+, O_2, NO, and sometimes OH. Some examples of the use of molecular emission used as a diagnostic in arc jets are given in the following subsections.

N_2^+(1st neg.) System

The $B^2\Sigma_u^+$-$X^2\Sigma_g^+$ transition of the N_2^+ molecule, is referred as the N_2^+(1st negative) system. One of the earliest examples of the rotational temperature determination in an arc jet flow is that of Shipley[21] who obtained the average rotational temperature in the jet free stream from the N_2^+(1st neg.) bands, (0,0) and (0,1). His Boltzmann plot is given in Fig. 18 for theR and P branches of the (0,0) band . Shipley was not able to use the technique in the shock layer of a blunt body because of other overlapping features of the spectrum and lack of spectral resolution.

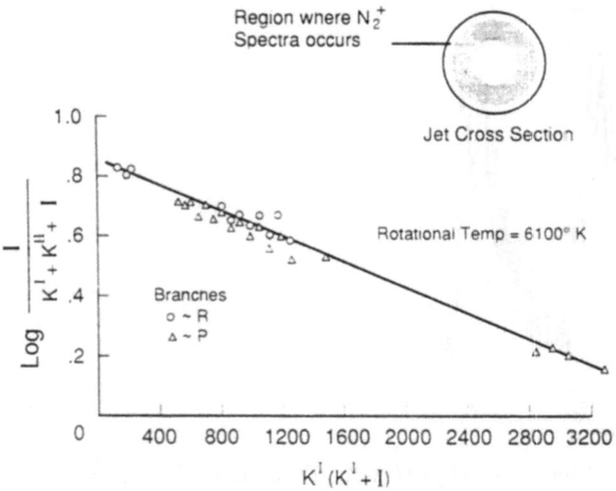

Fig. 18 Boltzmann plot of N_2^+ rotational lines from nitrogen arc jet. (From Shipley[21])

Park[30] obtained the vibrational temperature in the stagnation region of the expansion nozzle by comparing the measured N_2^+(1st neg.) spectrum with a calculated spectrum of Whiting, Arnold and Lyle.[37]

Recently the N_2^+(1st neg.) spectrum was used to determine the vibrational and rotational temperature. Blackwell, et al.[38] fit the (0,1) and (1,2) bands to a computed spectrum[39,40] to obtain the rotational and vibrational temperature distribution in the shock layer of a nitrogen arc jet at the Johnson Space Center. Examples of the spectra are shown in Fig. 19 and the resulting temperature profile is shown in Fig. 20. It can

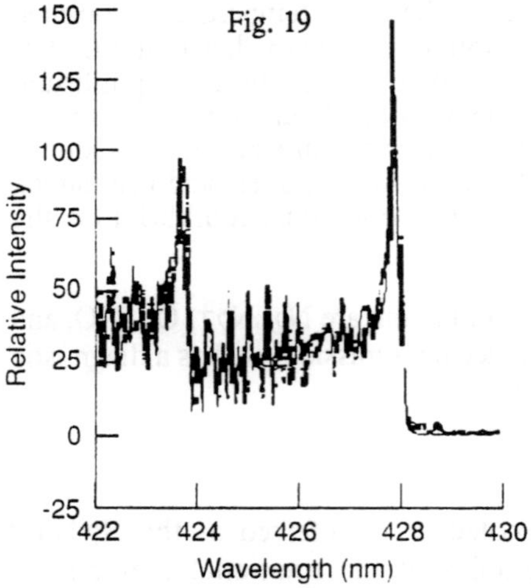

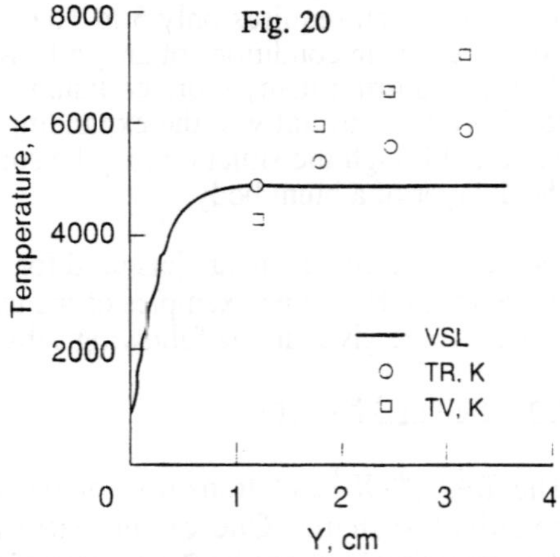

Fig. 19 Measured and computed spectra showing N_2^+(1st neg.) (0,1) and (1,2) bands at 1.27 cm from surface. (From Blackwell, et al.[38])

Fig. 20 Measured vibrational and rotational temperatures determined by spectral fitting compared with viscous shock layer single temperature calculation. (From Blackwell, et al.[38])

be seen that the two temperatures are not equal, implying that there has not been enough time for the temperatures to relax to a common equilibrium temperature. The computed temperature in Fig. 20 was based on a single temperature gas model and cannot discriminate thermal nonequilibrium effects. The method of comparing the measured and computed spectra is required here because the resolution of the spectrometer is not sufficient to resolve the rotational structure in which the R and P branches overlap.

The existence of a Boltzmann distribution of vibrational states was also checked by Blackwell, et al.[38] The fits made by varying the relative population of the upper vibrational state indicated that the vibrational states of the N_2^+(1st neg.) system deviate from equilibrium and that a vibrational temperature cannot be defined at all locations in the shock layer. However, there was some evidence that overlapping bands may have caused some of this deviation.

N_2(1st pos.) System

A more difficult system to analyze than N_2^+ is the N_2(1st positive) system. Its transitions are between the $B^3\Pi_g$ and $A^3\Sigma_u^+$ states for which there are many overlapping vibrational and rotational bands. Fig. 17 shows the system, where the broad peaks approximately correspond to regions of constant $\Delta v = v''-v'$. Without an extremely high resolution spectrum the only way to extract temperature information from this spectrum is to fit a computed spectrum to the measured spectrum. Blackwell, et al.[41] compared the computed ratios of integrals of intensities in spectral regions with measured ratios to obtain vibrational and rotational temperatures. This procedure was applied to radiation from the shock layer of a blunt body in an arc jet air flow.

OH Spectrum

Water vapor from steam ejector vacuum systems and desorbed from surfaces sometimes contaminate the flow in arc jets. Also OH may be produced in boundary layers around ablating test articles. Radiation form OH can be used as a diagnostic. Fishburne and Petrie[42] measured the rotational temperature distribution across an arc jet nozzle flow using OH as a diagnostic. They presumed that the OH is present as a trace species in the argon gas supply or is from minute leaks in the test chamber. To determine the rotational temperature Fishburne and Petrie developed Boltzmann plots of the Q_1 branch of the transition $^2\Sigma$-$^2\Pi(0,0)$. Their temperature distributions are shown in Fig. 21, which also shows the total temperature and pressure in the arc heater upstream of the nozzle throat. These results indicate that the flow is fairly uniform, but peaked slightly in the center of the jet.

3.2.1.3.5 Line Absorption

Many species of interest in arc jet flows do not radiate. For example ground state atoms and molecules cannot radiate and the transition probabilities of atoms and molecules in metastable states is zero to very low. Therefore, these species do not emit appreciably. To determine the population of these ground and metastable states one can do absorption measurements. The transition probability for a transition from a lower state to an upper state is uniquely related to the transition from the upper to the lower state. Therefore, by illuminating the flow with radiation from a suitable source and measuring its attenuation through the gas one can obtain a measure of the population of the lower state. Such an experiment was carried out by Valentin, et al.[43]

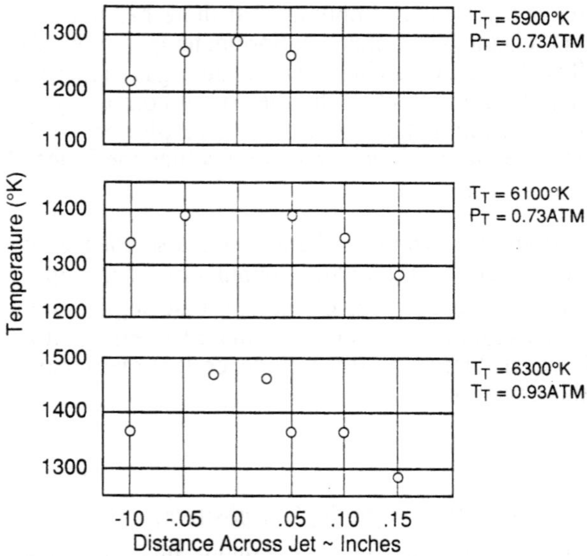

Fig. 21 OH rotational temperature distribution across core of exhaust jet. (From Fishburne and Petrie[42])

in an argon arc jet flow. They measured the metastable population distribution in the thermal boundary layer of a metallic flat plate whose surface was aligned parallel to the flow. Their metastable density distribution is shown in Fig. 22. They also measured the population of electrons and the electron temperature distributions from the emission line intensities. They found that the population of the metastable and the other excited states was much larger than expected from equilibrium considerations. The excitation temperature profile in the boundary layer seen in Fig. 23 showed an unexpected increase near the wall when the boundary layer edge temperature was low. They postulated that the increase was due to ions colliding with the wall causing Auger electrons to be ejected from the surface with an energy of a few electron volts, which heated the boundary layer.

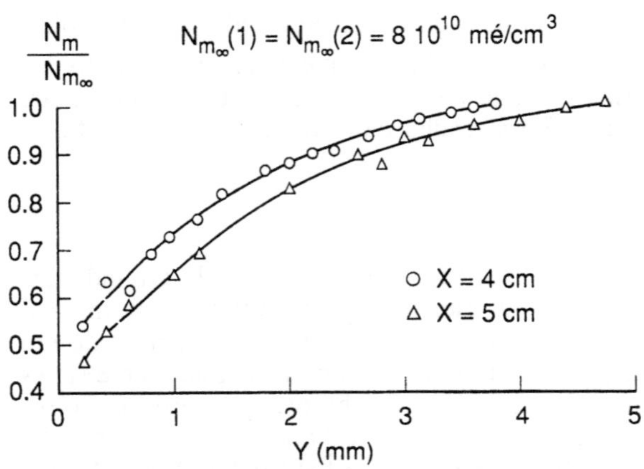

Fig. 22 Argon metastable density profiles in the boundary layer of a flat plate. (From Valentin et al.[43])

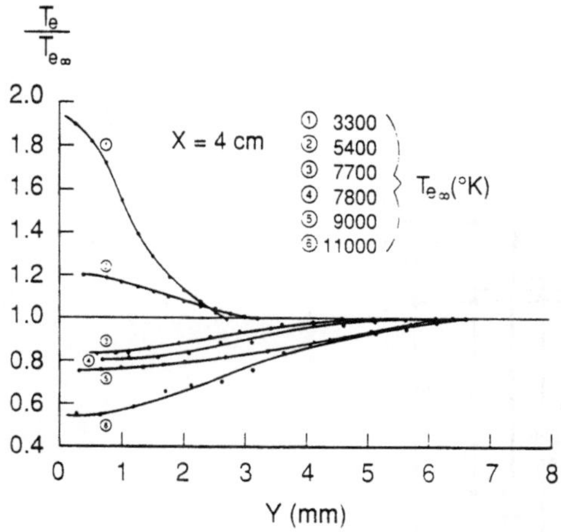

Fig. 23 Electron temperature profiles in the boundary layer of a flat plate with various freestream temperature conditions. (From Valentin et al.[43])

3.2.2 Laser Scattering

Much information about the flow is available from emission spectra, However, the measurements are sums of the radiation along the line of sight of the spectrometer. If the flow is not uniform then the measurement results in an average over the line of sight, this is not a local quantity. To obtain local information from line of sight measurements of axisymmetric flows one may use the Abel inversion technique. However, this technique is cumbersome and introduces some error into the results. Also there are many cases in which the flow does not have axial symmetry. Laser scattering can overcome the problem of line of sight averages because the laser illumination can be focused along one line while the observation or measurement is made along a line orthogonal to the laser. Thus the volume of gas formed by the intersection of the laser beam and the field of view is very small.

There are several types of laser scattering techniques that have been used to measure properties in gases. These include Rayleigh scattering, stimulated Raman scattering, coherent antistokes Raman scattering (CARS and BOXCARS), Mie scattering, laser induced fluorescence (LIF), and Thomson scattering. Of these techniques only the last two have as yet been applied to arc jet flows. LIF and Thomson scattering will be discussed in the following paragraphs.

3.2.2.1 Laser Induced Fluorescence

When molecules or atoms are illuminated with radiation whose frequency corresponds to a transition between states, the species may absorb the radiation and subsequently emit the energy to allowed lower states. Examples of the such transitions are shown in Fig. 24. The rate at which the upper stated is formed depends on the density of molecules in the lower state N_l, the absorption transition probability A_{lu}, and the intensity of the laser I_L, and the energy per photon $h\nu_{lu}$

$$\frac{dN_u}{dt} \propto \frac{N_l \, I_L \, A_{lu}}{h\nu_{lu}} \tag{19}$$

As in the case of spontaneous emission the amount of radiation that is emitted from the excited molecule is proportional to the emission transition probability A_{uj} from upper

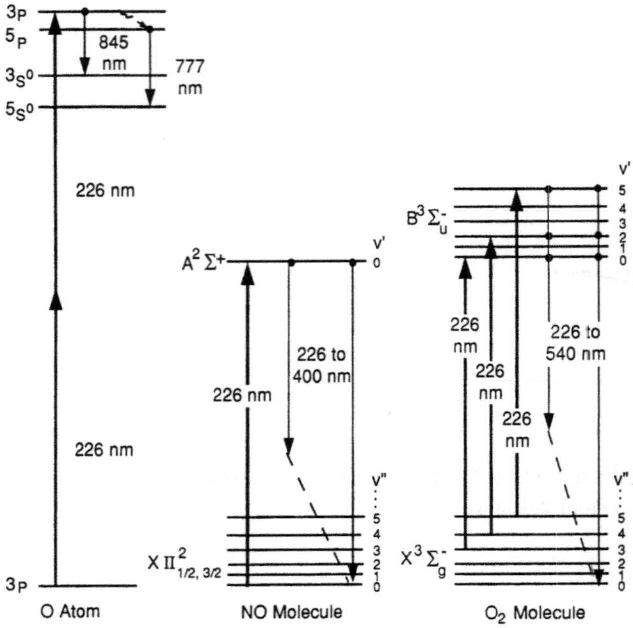

Fig. 24 Energy level diagrams for O, NO, and O_2. (From Arepalli[44])

state u to each lower state j. The loss of upper state molecules due to radiation and collisions is given by

$$\frac{dN_u}{dt} = \sum_j N_u A_{uj} + N_u f_u \tag{20}$$

where N_u is the number of molecules in the upper state and f_u is the collision frequency for deexcitation. Equations (19) and (20) can be solved to determine various properties of the flow, depending what is known. If the only source of excitation is the laser radiation and the radiative lifetime of the excited state is short compared with the collisional lifetime ($A_{uj} >> f_u$), then the measured intensity of the emitted radiation will yield the density of the lower state. In the case of molecules, the laser wavelength can be tuned so that the density of the lower states is determined. Thus, as in the case of emission, the excitation temperature of the molecules lower states can be obtained, now without the complication of overlapping bands, and with the advantage that the measurement is local.

A variety of wavelength ranges is possible using various types of lasers and dyes. Thus LIF access is available for diagnosing many molecules and transitions.

Arepalli[44] demonstrated the ability to obtain meaningful LIF signals from NO and O_2 in the free stream and shock layer of an arc jet at NASA Johnson Space Center. He produced a tunable (approximately 226 nm) ultraviolet laser signal using the 355 nm wavelength line of the pulsed YAG laser to pump a tunable dye laser whose output was about 430 to 470 nm. The wavelength of the output beam was halved to about 225 nm using a crystal doubler. The output power in this range was about 1.2 mJ and the bandwidth was 0.01 nm. The LIF signal from the flow zone was measured with a photon counting photomultiplier gated to reduce background radiation from the flow. Filters in front of the photomultiplier restricted the spectrum to pass only light from

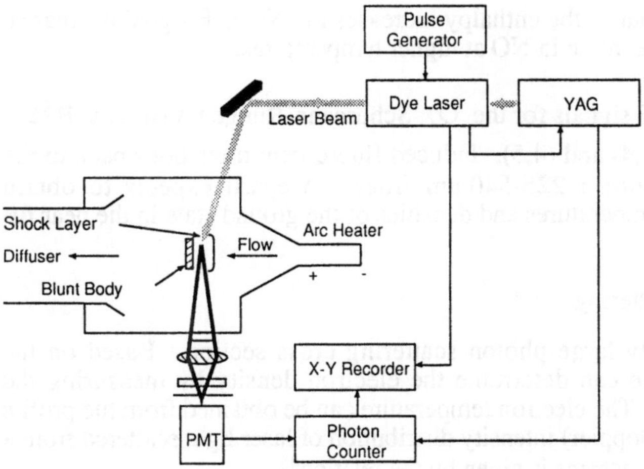

Fig. 25 LIF experimental set up. (From Arepalli[44])

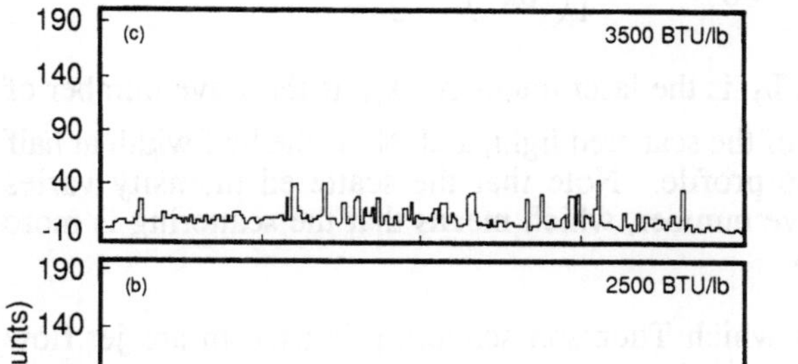

Fig. 26 LIF spectrum of (0,0) band of NO(A-X) in free stream of arc jet under three enthalpy conditions. (From Arepalli[44])

specific species. The optical arrangement is shown in Fig. 25. Preliminary results indicate that at low enthalpy NO and O_2 are readily detectable. The energy levels and the transitions studied are shown in Fig. 24. By tuning the output of the laser the various ground electronic state vibrational levels $X^2\Pi_{1/2,3/2}$ could be excited to the upper states $A^2\Sigma^+$. The fluorescent signals from the upper states were then measured with a wide band photomultiplier. Sample NO spectra are shown in Fig. 26 obtained in the free stream with filters in place to detect the NO $A^2\Sigma^+$—$X^2\Pi_{1/2,3/2}$. It can be seen from the three scans that as the enthalpy increases the NO LIF signal decreases. This is consistent with the decrease in NO at higher temperatures.

Arepalli[44] also obtained LIF signals for the O_2 Schumann-Runge transitions $B^3\Sigma_u^-$ —$X^3\Sigma_g^-$ (v',v") = (0,3), (2,4) and (4,5). Induced fluorescent transitions back to the X state were measured with a 225-540 nm filter. Arepalli expects to obtain vibrational and rotational temperatures and densities of the ground state in the near future from LIF spectra.

3.2.2.2 Laser Thomson Scattering

Free electrons have a fairly large photon scattering cross section. Based on the scattering cross section one can determine the electron density by measuring the absolute scattered intensity. The electron temperature can be obtained from the profile of the scattered line. The (Doppler) intensity distribution of laser light scattered from a Maxwellian distribution of electrons is given by the relation[45]

$$I(k) = \frac{N_e r_e I_o \sin\theta}{k_o^3} \exp\left[\left(\frac{k-k_o}{\delta k}\right)^2 \ln 2\right] \qquad (21)$$

where r_e is the electron radius, I_o is the laser intensity, k_o is the wave number of the laser, k is the wave number of the scattered light, and δk is the half width at half maximum of the measured line profile. Note that the scattered intensity varies inversely as the cube of the wave number, which means that the scattering is more efficient at long laser wavelength.

The only known experiment in which Thomson scattering is used in arc jet flow diagnostics is that of Michels et al.[46] in which they measured temporal and radial profiles of electron density and temperature in a pulsed magneto-plasmadynamic arc thruster. Their apparatus is shown in Fig. 27. They found that the average density (in the range of about 10^{14} cm^{-3}) is rather constant over 100 µsec, but fluctuated significantly. Likewise the average electron temperature, which was in the range of about 4 eV was also fairly constant. The radial profiles were relatively flat at 4 cm from the nozzle exit.

Thomson scattering, most useful at high N_e, requires that the background scattered light be eliminated with the use of beam dumps and baffling. The Rayleigh scattered line, which appears at the center wavelength of the Thomson scattered line, must also be filtered out by efficient notch filters so as not to allow significant background light into the detection spectrometer.

3.2.2.3 Laser Raman Scattering

Laser Raman scattering involves a scattering light from species that results in a frequency shift of the scattered light relative to the frequency of the incident beam. The difference in frequency depends on the energy states of the scattering species. Therefore, the spectrum of the scattered light is determined by the population of states in the scattering molecule. It can probe the spectrum of states inaccessable to the emission, resulting in a detectable scattered signal that has the spectral signature of, say the ground electronic state molecule. Boiarski and Daum[47] have studied the feasiblity of using laser Raman scattering for measuring number densities and vibrational temperature of nitrogen, oxygen and nitric oxide. Their exploratory experiments

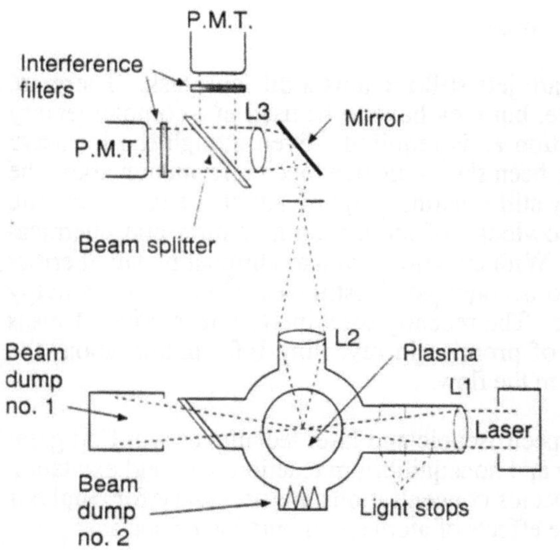

Fig. 27 Schematic of Thomson scattering diagnostic equipment. (From Michels et al.[46])

in the Air Force Flight Dynamics Laboratory 50-Megawatt Reentry Nose Tip (RENT) Facility have showed the feasibility of the technique for arc tunnel flow diagnostics.

3.2.3 Microwave Attenuation and Interferometry

In the interest of completeness microwave diagnostics will be mentioned briefly. However, details will be left for the reader to uncover. Suffice it to say that microwave propagation through an ionized medium is affected by the free electrons, which result in a phase shift of the wave. At sufficiently high densities, the wave is also attenuated. If fact at the critical cutoff frequency microwaves will not propagate at all. Microwave diagnostics was used by Duckett and Jones[13] in their arc jet electron density measurements in which they got fair agreement with Langmuir probe measurements.

3.2.4 Photographic Flow Visualization

Photography is a common method of determining qualitative characteristics of arc jet flows. Both still and motion pictures can yield information about the flow, especially when something unexpected goes wrong with the facility or the test article. One can see when melting occurs or if the heating is nonuniform. Often a qualitative heat flux distribution is seen in the photographs of test articles, for example, noting the location of a shock or vortex impingement. Since the flow field is self luminous one of the frequent uses for photography is to determine the shock shape around test articles and to determine where there are unwanted shock waves such as shock diamonds or Mach disks in the free stream flow.

Quantitative information can be obtained about the Mach number by measuring the Mach angle of an oblique shock around a wedge. Park and Okuno[48] measured the shock angles from photographs of the luminous shock about a 15° wedge in an arc jet flow. They determined that the velocity was 10.7 km/s ±3% based on a total flow enthalpy of 170 MJ/kg. The uncertainty is due to their not measuring the static pressure of the flow. They also made time of flight measurements which confirmed the results. One problem with shock angle measurements is the low density in arc jets which causes shock curvature via boundary layer interaction.

4. Conclusions

Determining the flow properties in arc jets still remains a difficult task. There are many diagnostic techniques available, but they have to be used in a complementary way to determine as much information as is required. Even though the intrusive techniques disturb the flow they have been shown to give much information about the flow. Determining the total enthalpy still remains a significant challenge. Accurate heat flux measurements require knowledge of atom recombination and chemical energy accommodation coefficients. With attention given to eliminating small errors and taking all pertinent physics into account good estimates of the total enthalpy distribution in the flow can be made. The recently used modern technique of mass spectrometry appears to hold a lot of promise in revealing information about the species and chemical nonequilibrium in the flow.

State specific methods, such as the spectroscopic and laser techniques will find great utility in understanding the chemistry and nonequilibrium reaction rates and excitation rates in the flow. These rates and species concentrations are necessary for catalysis measurements and for determining the effects of atom recombination on surfaces.

Little used methods such as the laser Thomson scattering and microwave techniques have been included for completeness. These techniques may find more use at very high enthalpy, in flows required to simulate very high speed reentries such as those associated with Lunar or Mars missions.

Finally photography is seen to be very useful in determining qualitative information about the flow and the test articles. Even photographic techniques may be useful in quantitative estimates of the velocity of the flow and perhaps validation of the flow chemistry via shock standoff distance measurements.

REFERENCES

[1]Pope, R. B., "Measurements of Enthalpy in Low-Density Arc-Heated Flows," *AIAA Journal,* Vol. 6, Jan. 1968, pp. 103-110.

[2]Winnovich, W., "On the Equilibrium Sonic-flow Method for Atmospheric Re-entry Research," NASA TN D-2132, 1964.

[3]Shapiro, A. H., *The Dynamics and Thermodynamics of Compressible Fluid Flow, Vol. 1,* The Ronald Press, New York, 1953.

[4]Hiester, N. K. and C. F. Clark, "Feasibility of Standard Evaluation of Procedures for Ablating Materials," NASA CR-379, 1966.

[5]Fay, J. A. and F. R. Riddell, "Theory of Stagnation Point Heat-transfer in Dissociated Air," *J. Aerospace Sci.*, Vol. 25, pp. 73-85, 1958.

[6]Bade, W. L. and J. M. Yos, "The NATA Code — Theory and Analysis," NASA CR-2547, 1975.

[7]Boison, J. C. and H. A. Curtiss, "An Experimental Investigation of Blunt Body Stagnation Point Velocity Gradient," *ARS Journal*, Vol. 29, pp. 130-135, 1959.

[8]Scott, C. D., "Catalytic Recombination of Nitrogen and Oxygen on High Temperature Reusable Surface Insulation," *Progress in Astronautics and Aeronautics,* Vol. 77, edited by A. L. Crosbie, 1981.

[9]Scott, C. D., "Measured Catalycities of Various Candidate Space Shuttle Thermal Protection System Coatings at Low Temperatures," NASA TN D-7113, 1973.

[10]Scott, C. D., "An Experimental and Analytical Study of Slip and Catalytic Boundary Conditions Applied to Spheres in Low Reynolds Number Arc Jet Flows," *Rarefied Gas Dynamics*, Proceedings of the Ninth International Symposium, edited by M. Becker and M. Fiebig, pp. D.14-1—D.14-12, 1974.

[11]Li, C.-P., "Numerical Solution of Viscous Reacting Blunt Body Flows of a Multicomponent Mixture," AIAA Paper 73-202, 1973.

[12]Grey, J., *Rev. Sci. Inst.*, Vol. 34, No. 8, Aug. 1963, pp. 857-859.

[13]Duckett, R. J., and W. L. Jones, Jr., "Free-stream Electron Concentration in an Arc Heated Wind Tunnel and Correlation of Langmuir Probe and Microwave Interferometer Measurements," NASA TN D-5500, 1969.

[14]Kamimoto, G., and M. Nishida, "Langmuir Probe Measurements in an Argon Plasma Flow," Dept. of Aeronautical Engineering Kyoto University Current Papers, CP. 8, Nov. 1965.

[15]Poissant, G., and M. Dudeck, "Velocity Profiles in a Rarefied Argon Plasma Stream by Crossed Electrostatic Probes," *J. Applied Phys.*, Vol. 58, pp. 1772-1779, 1985.

[16]Founoune, M., C. Danton, and M. Dudeck, "Stationary Free Jets of Nitrogen Plasma," Proceedings of the 15th International Symposium on Rarefied Gas Dynamics, Grado, Italy, 1988.

[17]Lasgorceix, P., M. Dudeck, and J. Caressa, "Measurements in Low Pressure, High Temperature and Reactive Nitrogen Jets," AIAA Paper 89-1919, 1989.

[18]Willey, R. J., "Direct Measurements of Atomic and Molecular Concentrations in Boundary and Surface Layers of Shocks," Northeastern University, Final report Grant No. NAG 9-260, March 13, 1989.

[19]Willey, R. J., "Gas Composition Measurements in Arc Jet Tunnels via Mass Spectrometry," to be published in *J. Thermophysics and Heat Transfer*.

[20]Scott, C. D., "Line Broadening of the Triplet Diffuse Series of Helium in an Arc-Jet Plasma," NASA TM X-58050, 1970.

[21]Shipley, K. L., Some Spectroscopic Measurements of a Supersonic Plasmajet," *Applied Spectroscopy*, Vol. 20, 1966, pp. 6-11.

[22]Gatterer, A., "Grating Spectrum of Iron," Vatican Press, 1951.

[23]Stair, R., R. G. Johnston, and E. W. Halbach, "Standard of Spectral Radiance for the Region of 0.25 to 2.6 Microns," *J. of Research of the National Bureau of Standards—A. Physics and Chemistry,* Vol.64A, 1960, pp. 291-296.

[24]MacPherson, H. G., "The Carbon Arc as a Radiation Standard," *J. Optical Soc. of America,* Vol. 30, 1940, pp. 189-194.

[25]Null, M. R., and W. W. Lozier, "Carbon Arc as a Radiation Standard," *J. Optical Soc. of America,* Vol. 52, 1962, pp. 1156-1162.

[26]Flemming, J. C., "An Evaluation of an High Temperature Blackbody as a Working Standard of Spectral Radiance," *Applied Optics,* Vol. 5, 1966, pp. 195-200.

[27]Bowen, S. W., "Spectroscopic and Optical Studies of a High Pressure, Under expanded Jet," AAIA Paper 66-164, 1966.

[28]Adcock, B. D., "Temperature Measurements on a Subatmospheric Argon Plasma Jet," *J. Quant. Spesc. and Radiat. Transfer,* Vol. 7, 1967, pp. 385-400.

[29]Greenshields, D. H., "Spectrometric measurements of gas temperatures in arc heated jets and tunnels," NASA TN D-1960, 1963.

[30]Park, C., "Comparison of Electron and Electronic Temperatures in Recombining Nozzle Flow of Ionized Nitrogen-Hydrogen Mixture. Part 2. Experiment," *J.Plasma Physics,* Vol. 9, part 2, 1973, pp. 217-234.

[31]Green, L. C., N. C. Johnson, and E. K. Kolchin, "Oscillator Strengths for Singlet and Triplet Series of Neutral Helium," *Astrophysical J.* , Vol. 144, 1966, pp. 369-375.

[32]Martin, W. C., "Energy Levels and Spectrum of Neutral Helium (^{4}He I)," *J. Research NBS, Section A (Phys-Chem),* Vol. 64, 1960, pp. 19-28.

[33]Griem, H. R., "Plasma Spectroscopy," McGraw-Hill, New York, 1964.

[34]Park, C., "Comparison of Electron and Electronic Temperatures in Recombining Nozzle Flow of Ionized Nitrogen-Hydrogen Mixture. Part 1 - Theory," *J.Plasma Physics,* Vol. 9, part 2, 1973, pp. 187-215.

[35]Anderson, A. D. and H. R. Griem, "Continuum Emission Coefficients form the Quantum Defect Method," Proceedings of the 6th International Conference on Ionization Phenomena in Gases, North Holland Publishing Co., Amsterdam, Vol. 3, 1963, pp. 293-295.

[36]Herzberg, G., *Molecular Spectra and Molecular Structure, I. Spectra of Diatomic Molecules, 2nd Ed.,* D. Van Nostrand Co., Princeton, NJ, 1950.

[37]Whiting, E. E., J. O. Arnold, and G. C. Lyle, NASA TN D-5088, 1964.

[38]Blackwell, H. E., E. Yuen, S. Arepalli, and C. D. Scott, "Nonequilibrium Shock Layer Temperature Profiles from Arc Jet Radiation Measurements," AIAA -89-1679, 1989.

[39]Park, C., "Nonequilibrium Air Radiation (NEQAIR) Program: Users Manual," NASA TM-86707.

[40]Park, C., "Calculation of Nonequilibrium Radiation in the Flight Regimes of Aeroassisted Orbital Transfer Vehicles," in Thermal Design of Aeroassisted Orbital Transfer Vehicles, H. F. Nelson, ed., *Progress in Astronautics and Aeronautics*, Vol. 96, 1985, pp. 395-418.

[41]Blackwell, H. E., Wierum, F. A., and Scott, C. D., "Spectral Determination of Nitrogen Vibrational Temperatures," AIAA-87-1532, 1987.

[42]Fishburne, E. S. and Petrie, S. L., "Spectrographic Analysis of Plasma Flows," Ohio State University Research Foundation, Columbus, OH, Rept. ASD-TDR-63-98, Contract No. AF 33(657)-7561, 1963.

[43]Valentin, P., Terrier, M., Vervisch, P. and Piar, G., "Laminar Boundary Layers in Low Pressure Argon Plasma," in *Dynamics of Ionized Gases*, Lighthill, M. J., Imai, I., and Sato, H., eds., John Wiley & Sons, 1971.

[44]Arepalli, S., "Demonstration of the Feasibility of Laser Induced Fluorescence for Arc Jet Flow Diagnostics," NASA CR-185595, 1989.

[45]Brown, T. S. and D. J. Rose, "Plasma Diagnostics using Lasers: Relations between Scattered Spectrum and Electron-Velocity Distribution," *J. Applied Physics,* Vol. 37, 1966, pp. 2709-2714.

[46]Michels, C. J., Rose, J. R., and Sigman, D. R., "Temporal Survey of Electron Number Density and Electron Temperature i the Exhaust of a Megawatt MPD - Arc Thruster," AIAA 72-209, 1972.

[47]Boiarski, A. A. and Daum, F. L., "An Application of Laser-Raman-Spectroscopy to Thermochemical Measurements in an Arc-Heated Wind Tunnel Flow," ARL-72-0126, 1973.

[48]Park, C. and Okuno, A. F., "Diagnosis of High-Density, Highly Ionized Nitrogen Wind Tunnel Flows," in proceedings of the 2nd International Congress on Instrumentation in Aerospace Simulation Facilities, P. L. Clemens, ed., Stanford University, Aug. 29-31, 1966.

CFD FOR HYPERSONIC PROPULSION

Louis A. Povinelli*

National Aeronautics and Space Administration
Lewis Research Center
Cleveland, Ohio 44135

Introduction

This paper presents some research activity on the application of CFD for hypersonic propulsion systems, which the author presented at the workshop. Since this author was requested, on one days notice, to fill in for an invited speaker who was unable to attend, the material contained herein is more representative of what was in hand rather than the total CFD effort that is underway at the authors laboratory and elsewhere. The presentation addressed the following items:

- o Propulsion system integration
- o Typical computations for propulsion components
- o CFD validation issues
- o Prognoses for success

To a large extent, the comments and illustrations used herein are based on a presentation made by the author at the Ninth ISABE meeting in Greece in September 1989 (ref. 1).

---

*Deputy Chief, Internal Fluid Mechanics Division.

Propulsion System Integration

In the mid 1970's, propulsion testing of a hypersonic ramjet engine (HRE) was performed at Mach 5 to 7 at the Plum Brook Station of the NASA Lewis Research Center. That configuration was axisymmetric in design and had a rather small annular passage through the combustor. The HRE was representative of a pod-mounted system rather than the highly blended configurations of today. In this presentation, the ability of CFD codes to simulate propulsion system components is discussed relative to the integrated engine body configurations which are more typical of today's designs. A generic version of such a highly blended configuration is shown in figure 1.

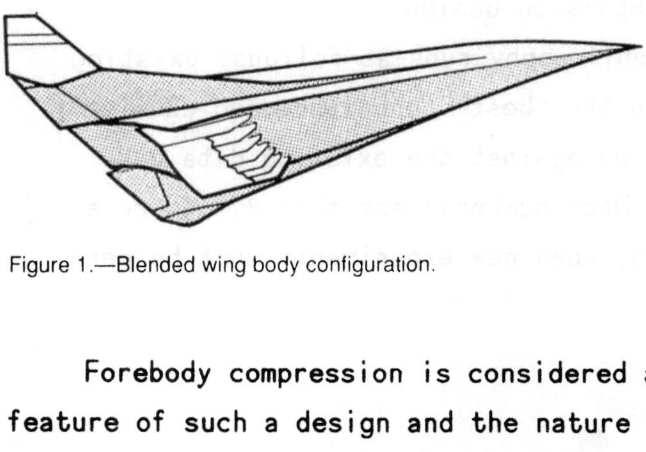

Figure 1.—Blended wing body configuration.

Forebody compression is considered an essential feature of such a design and the nature of the corresponding boundary layer must be taken into consideration for inlet behavior. Needless to say, a great deal of vehicle research and testing is required for proper aerodynamics as well as being capable of delivering uniform flow to the inlet. The propulsion system is assumed to be a combined ramjet/scramjet system having a common flowpath.

Propulsion Modes: The common flowpath engine considered in this paper is envisioned to operate as a subsonic combustion ramjet over the flight Mach number from 3 to 6. At higher flight speeds, the supersonic combustor mode would be employed up to flight numbers which may be on the order of Mach 15. This upper limit is speculative and depends on a number of unknown factors. Above the upper limit of air-breathing operation, integrated rocket thrusters would be employed to achieve orbital velocities. The ramjet modes are illustrated in figure 2.

Hypersonic Propulsion Design Approach: Given the aircraft propulsion system illustrated in figure 2 and the operational modes for the engine, one may inquire how to approach its propulsion design.

The current philosophy runs as follows; existing computer codes with the "best" turbulence and chemistry modeling are assessed against the existing data base which is mostly at Mach numbers less than 8. Where a lack of data exists, then new experiments must be per-

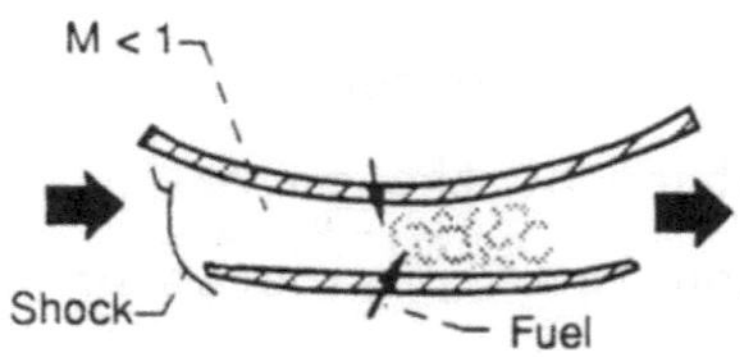

(a) Subsonic combustion ramjet.

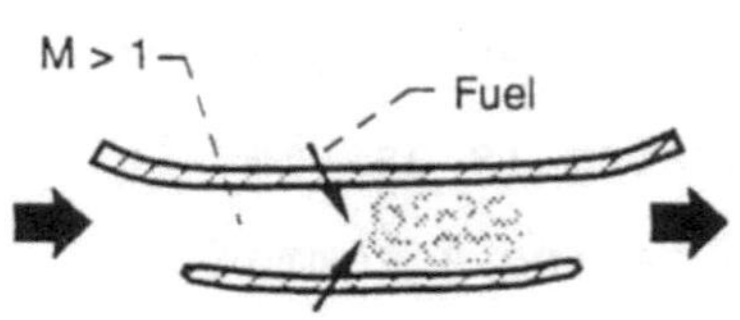

(b) Supersonic combustion ramjet.

Figure 2.—Ramjet operating modes.

formed. Numerous iterations between computations and experiments will eventually "validate" the codes. These validated codes, with all the sophistication of real gas effects and turbulence/chemistry closures will subsequently be extrapolated to the higher Mach numbers (e.g., M8 to M16) to assess various geometrical engine configurations. After a "sufficient" number of numerical computations, backed up by available pulse or shock facility data, a candidate design will emerge.

Flight experiments will then provide the next or "true" level of validation. Information from such testing will then be used to modify the physical and chemical modeling used in the simulations. As flight test speeds are increased incrementally over the Mach range required for orbit, the improved CFD simulations will provide guidance at each of the next incremental speed levels. Thus, flight testing and CFD simulations will be conducted "hand-in-hand" as hypersonic vehicles move up the speed corridor.

Above the upper limit of air-breathing operation, integrated rocket thrusters would be employed to achieve orbital velocities.

Typical Computations for System Components

Generic Inlet

The simple rectangular inlet configuration shown in figure 3 was tested at Mach 12.26. A flat plate of 30 in. length preceded the entrance to the inlet in order to simulate the boundary layer growth on the forebody of a hypersonic aircraft. Compression wedges form the top and bottom walls of the inlet and the contraction ratio

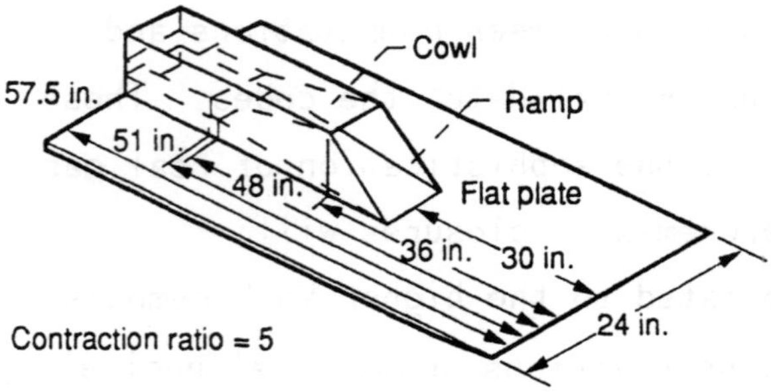

Figure 3. - Generic inlet (ref. 5).

was equal to five. Swept sidewalls which connect the upper and lower walls, prevent compressed flow from spilling over the inlet sidewalls.

Computations were made with a three-dimensional PNS LBI implicit scheme (ref. 2) with grids of 80 by 60 by 750 on a Cray X-MP. This solver includes real gas effects (ref. 3) as well as dissociation and ionization modeling (ref. 4). For this experiment, however, the inlet air was only heated sufficiently to avoid condensation, and the real gas modeling was not required. The issues that are of importance in this computation are the assumptions regarding the state of the boundary layer, the turbulence model, spillage of flow around the sideplates and shock boundary layer interaction. For the PNS computation it was assumed that the boundary layer was turbulent starting on the leading edge of the flat plate, the cowl leading edge and the sidewall leading edges. The turbulence model used was a Baldwin-Lomax model and spillage was not considered. Modeling of the shock boundary layer interaction involved the use of a flare approximation in order to allow the PNS to march through the region of flow separation. The results of the PNS solution are shown in figure 4. Contour plots of con-

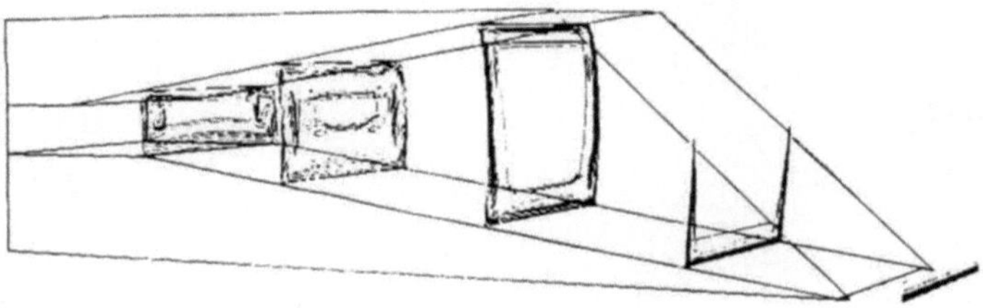

Figure 4.—Mach number contours, M = 12.25 (ref. 2).

stant Mach number within the inlet are shown. The concentration of lines near the walls indicate the boundary layers, while concentrated contours in the freestream, indicate shock wave locations. The flow features seen are boundary layer buildup on the flat plate followed by thickening on the sidewalls and ramp surface. Shocks generated by the compression wedges are seen as horizontal lines, and the sidewall shocks are vertical lines.

The low energy flow in the sidewall boundary layer has been swept up the sidewall by the ramp shock, and then down the sidewall by the cowl shock. Further downstream, the shock waves cross and are distorted by interaction with the sidewall boundary layers and the expansion fan on the ramp surface. Additional complex interactions then occur as the flow moves downstream. The PNS solution fails when the ramp shock wave reflects from the cowl and strikes the ramp surface, resulting in large corner separation of the low energy flow.

An alternate view of the three-dimensional flow is obtained with sidewall particle tracing (fig. 5). Interaction of the ramp and cowl shocks with the sidewall boundary layer causes the particles to converge near the shock interaction point. The particles are then displaced due to the vortex motion. Flow migration details

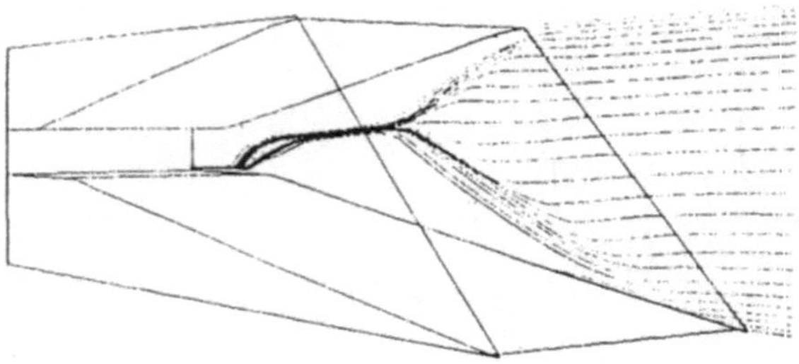

Figure 5.—Sidewall particle tracing, M = 12.25 (ref. 2).

are evident in this computational simulation. As a side-note, since the vortex persists downstream, it has been proposed that enhanced fuel mixing could occur with judicious injector locations downstream (ref. 1).

Navier-Stokes computations have also been carried out for the generic inlet at NASA Langley with CFL3D (ref. 5). In this case, the boundary layers were assumed turbulent on all surfaces from the leading edges. The turbulence model used was a Baldwin-Lomax model and spillage over the sideplates was not considered. In the vicinity of the shock boundary no special modeling was employed. Figure 6(a) shows the pressure distributions for the ramp and centerline cowl surfaces, using two different grids. Figure 6(b) shows the side plane distributions. Comparison of the CFL3D results and the experimental data show good agreement, particularly along the centerline where shock locations appear to be well resolved by the code. The viscous interactions occurring along the side plane are also accurately resolved. In addition, CFL3D was used to compute the heat transfer on the ramp and cowl surfaces (figs. 7(a) and (b)). The heat flux distributions are reasonably well predicted on both ramp and cowl surfaces.

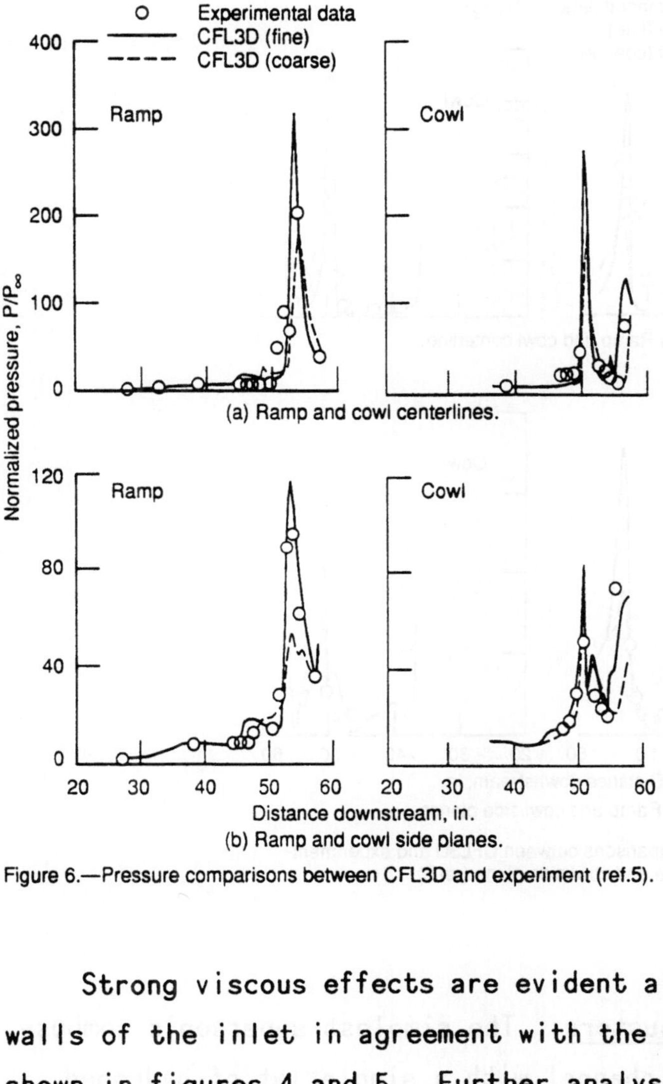

Figure 6.—Pressure comparisons between CFL3D and experiment (ref.5).

Strong viscous effects are evident along the side walls of the inlet in agreement with the complex behavior shown in figures 4 and 5. Further analysis of the Mach 12 inlet is underway at the NASA Centers and industry.

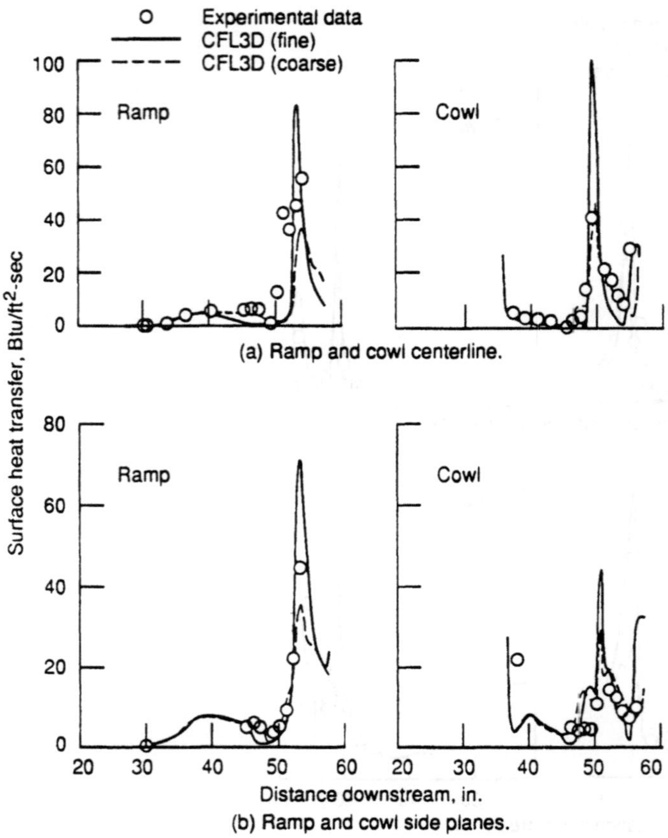

Figure 7.—Heat transfer comparisons between CFL3D and experiment-2-D model, ramp side plane and cowl side plane (ref.5).

Combustors

Simple Combustors: The simplest supersonic combustor scheme is a channel with a single jet of hydrogen injected normally to the supersonic stream, as illustrated in figure 8.

This reacting flow situation was solved using the RPLUS code at NASA Lewis, which is an LU algorithm. The grid used for the solution was 60X40X40 with grid clustering. The resulting Mach number distribution is shown in figure 9 with good fidelity and resolution of the injection fluid mechanics. Figure 10 shows the computed temperature contours by Drs. Yu and Shuen (ref. 6).

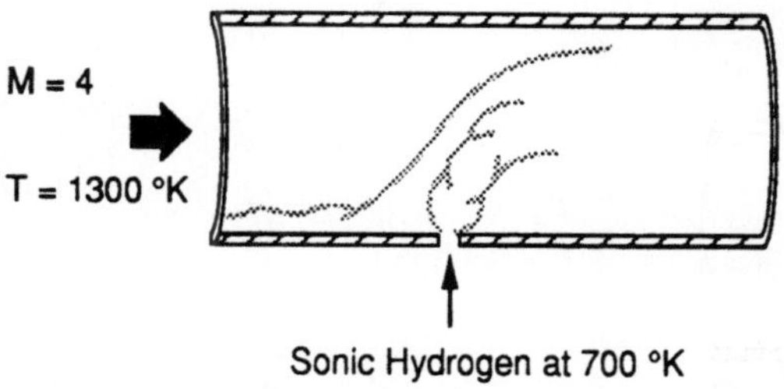

Figure 8.—Simple jet injection.

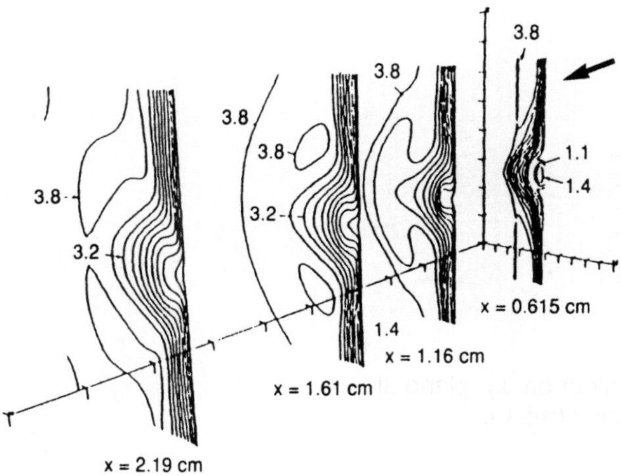

Figure 9.—Mach number contours on yz planes at various x locations for Case 1 (ref. 6).

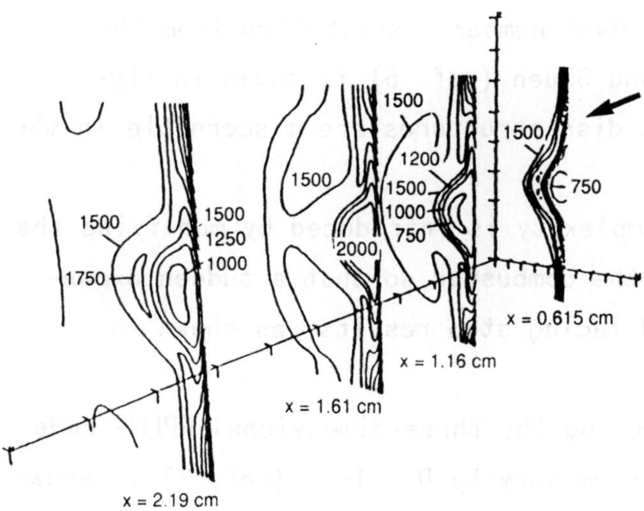

Figure 10.—Temperature contours on yz planes at various x locations for Case 1 (ref. 6).

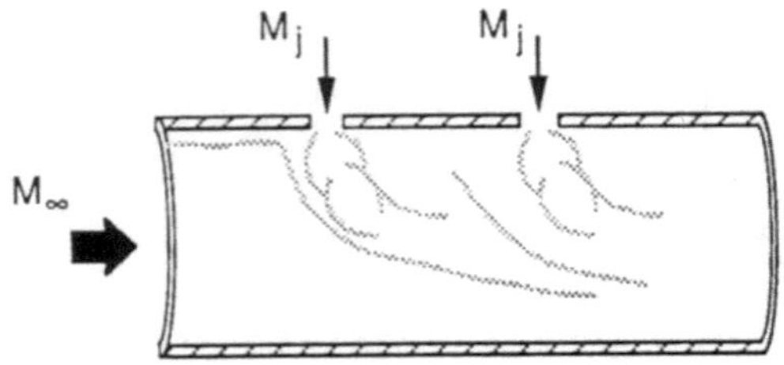

Figure 11.—Dual jet injection.

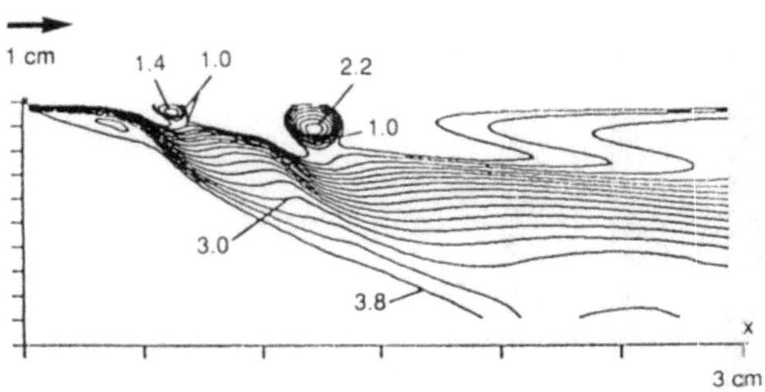

Figure 12.—Mach number contour on xy plane at center of injection port for Case 2 (ref. 6).

Dual Injection: A somewhat more complex injection scheme involves two jet injection ports which are aligned in the axial direction as illustrated in figure 11.

The resulting Mach number distribution from the RPLUS code by Yu and Shuen (ref. 6) is shown in figure 12. Both Mach disc structures are discernable in the computations.

Additional complexity is introduced by modifying the straight walls of the combustor so that a sudden expansion or a rearward facing step results, as shown in figure 13.

Computations using the three-dimensional RPLUS code with hydrogen-air chemistry by Dr. Tsai (ref. 7) is shown in figures 14(a) and (b). The temperature distributions appear reasonable. The results shown are laminar. Tur-

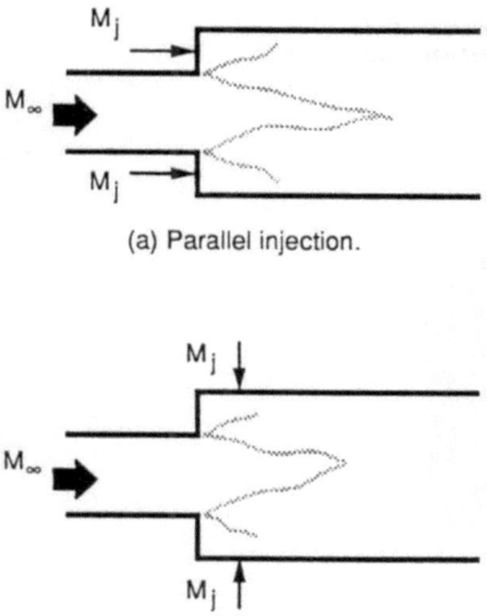

Figure 13.—Sudden expansion combustor.

bulence modeling needs to be incorporated into RPLUS for more realistic conditions. That activity is currently nearing completion.

Asymmetric Nozzle Geometry

A typical three-dimensional Navier-Stokes computation for a nonaxisymmetric nozzle is illustrated in figure 15. For this case, the supersonic jet is issuing into a quiescent atmosphere. A three-dimensional Navier-Stokes code, PARC, was used to study the flow behavior. A Baldwin-Lomax turbulence model was employed in the code. These computations were performed by Dr. Hen Lai of Sverdrup/Lewis (ref. 8). Figure 16 shows typical results of the spanwise variation of Mach number, starting at the nozzle center plane and extending to the side wall shear layers. Analyses of the type described above have been combined to provide a complete computation from vehicle nose through the propulsion system to the tail of the aircraft.

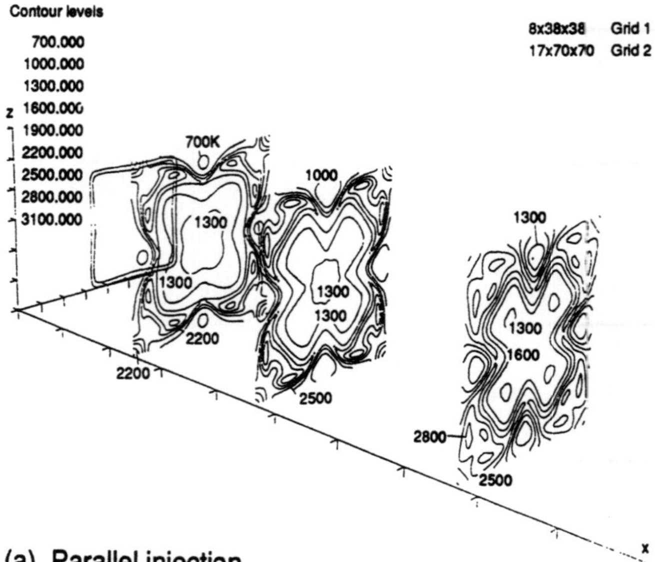

(a) Parallel injection.

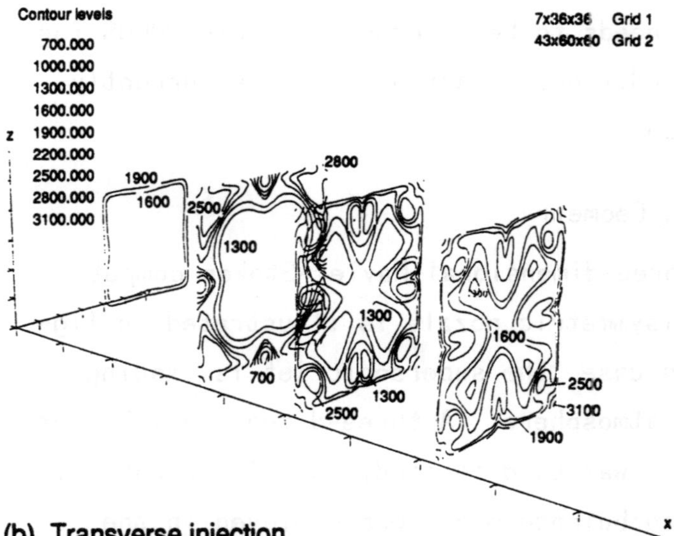

(b) Transverse injection

Figure 14.—RPLUS temperature computations for a sudden expansion combustor by Tsai (ref. 7).

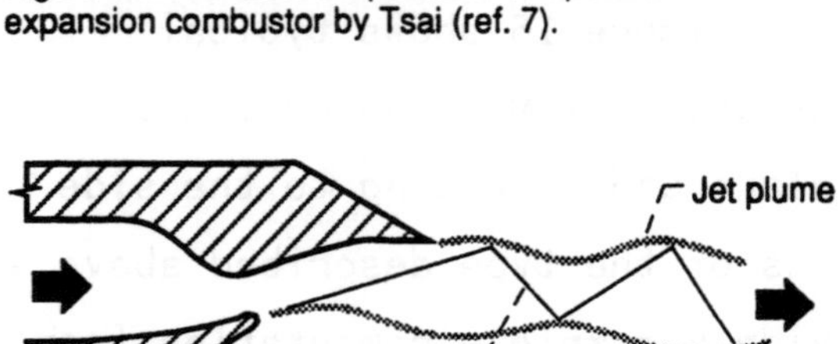

Figure 15.—Asymmetric nozzle configuration.

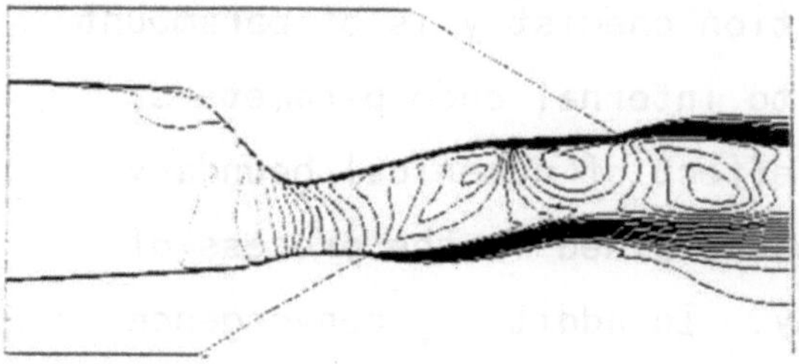

(a) Centerline.

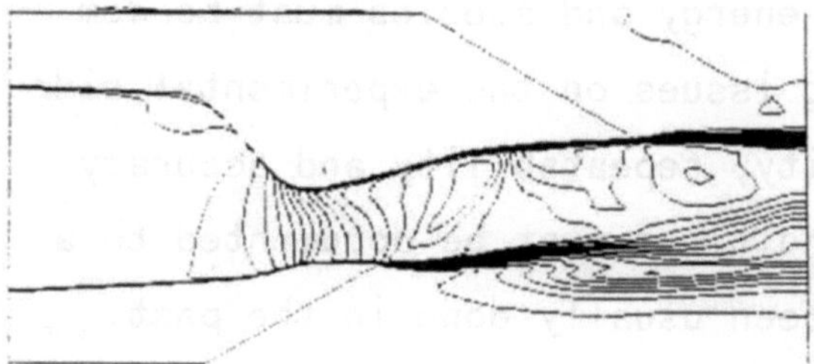

(b) Midpoint.

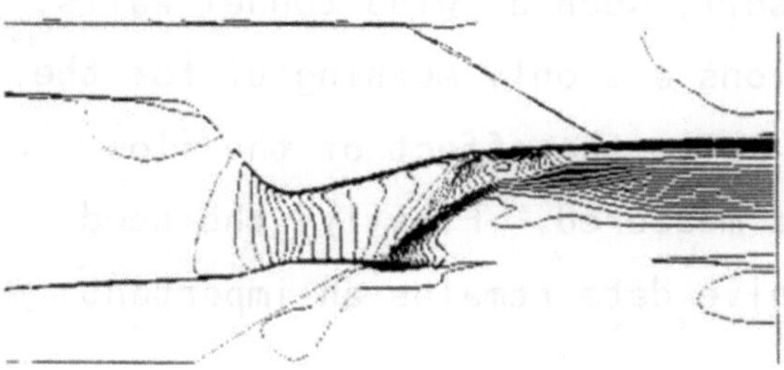

(c) Wall.

Figure 16.—Mach number contours, side views (ref. 8).

Validation Issues

Validation of numerical simulations must deal with a number of specific issues. In this section we will address:

- o General validation issues
- o Design issues
- o Critical research fora validation

General Issues: In this category, both computational and experimental issues need to be addressed. On the computational side, modeling of turbulence, boundary layer transition and reaction chemistry is of paramount importance. Sensitivity to internal code parameters, grid sensitivity and the effect of numerical boundary conditions must also be ascertained in the process of establishing code validity. In addition, convergence behavior; the ability to capture discontinuities and to preserve mass, momentum, energy and species must be demonstrated. Corresponding issues on the experimental side revolve around the validity, repeatability and accuracy of data. The initial conditions must be documented to a greater extent than has been usually done in the past. In addition, some testing is strongly affected by the experimental apparatus itself, such as wind tunnel walls, so that numerical simulations are only meaningful for the entire apparatus. At any rate, the effect of the flow facility must be known and measured. Finally, the need for fluctuating, nonintrusive data remains an important requirement.

Design Issues: As computer codes are put to use for "design" purposes, a number of critical questions arise. Some typical questions that have arisen are the following:

- To what extent are propulsion CFD codes validated?
- What is the degree of validation required?
- What are the propulsion design needs?
- What engineering parameters are needed?
- What computed variables are required to produce engineering parameters?
- To what extent are the computed variables affected by specific physical or chemical modeling?

Many discussions have ensured over the above questions, without a clear cut response to all of the inquiries. In general, one is led to the conclusion that a systematic study and resolution of all the computational and experimental validation issues raised in the previous section would require many years to complete. It appears prudent, therefore, at this time to develop fundamental understanding first, then to make judgements on the importance of various phenomena and lastly to perform numerical sensitivity studies.

Prognosis for Success

It is believed that modeling of the most important physics and chemistry will be improved only through a rigorous and systematic validation process. This process will involve a large number of experiments, from those with simple isolated phenomena to those involving multiple simultaneous effects. Corresponding modeling and code simulations must also be carried out in close concert with the measurements.

Furthermore, an improved understanding of the relevant flow physics and chemistry over a wide range of operating conditions will evolve incrementally from experimental flight data. These data will assist in the improvement of the required modeling for CFD codes and the codes, in turn, can be applied to the next increment in the vehicle's flight velocity.

Finally, it is believed that validated (which implies an important role of experiments) numerical simulations will play a major role in the design of future propulsion systems only if continuous effort is focused on the combined experimental/computational methodology.

References

1. Povinelli, L.A.: Advanced Computational Techniques for Hypersonic Propulsion. International Symposium on Air Breathing Engines (9th, 1989, Athens, Greece) F. Billing, ed., AIAA, Washington, DC, 1989, Vol. 2, pp. 993-1008. (Also, NASA TM-102005, 1989.)
2. Reddy, D.R., et al.: Three-Dimensional Viscous Analysis of a Hypersonic Inlet. AIAA Paper 89-0004, Jan. 1989. (Also, NASA TM-101474, 1989.)
3. Liou, M.-F.: Three-Dimensional PNS Solutions of Hypersonic Internal Flows With Equilibrium Chemistry. AIAA Paper 89,0002, Jan. 1989.
4. Yu, S.-T., et al.: Numerical Simulation of Hypersonic Inlet Flows With Equilibrium or Finite Rate Chemistry, AIAA Paper 88-0273, Jan. 1988.
5. Diley, A.; Eppart, W.; and Switzer, G.: Zonal Analysis of Two High Speed Inlets. Proceedings of the CFD Symposium on Aeropropulsion, NASA CP-10045, 1990, pp. 29-1 to 29-13.
6. Yu, S.-T.; Tsai, P.; and Shuen, J.-S.: Three-Dimensional Calculation of Subsonic Reacting Flows Using an LV Scheme. AIAA Paper 89-0391, Jan. 1989.
7. Lai, H.: Three-Dimensional of Single-Expansion-Ramp and Scramjet Nozzles. Proceedings of the CFD Symposium on Aeropropulsion, NASA CP-10045, 1990, pp. 32-1 to 32-24.
8. Tsai, Y.L.: Recent Update of the RPLUS2D/3D Codes. Proceedings of the CFD Symposium on Aeropropulsion, NASA CP-10045, 1990, pp. 34-1 to 34-12.

Experimental Simulation of High-Enthalpy Real-Gas Effects

Hans Hornung

Graduate Aeronautical Laboratories
California Institute of Technology
Pasadena, CA 91125

Abstract

This paper outlines the motivation for high enthalpy gas dynamics research and the rationale leading to the free-piston driver technique in relation to other flow simulation techniques and to flight tests. It discusses selected important technical and scientific problems demanding fundamental research. The philosophy of the approach to this research, involving the application of modern diagnostic techniques and close coupling with computational fluid dynamics, as well as direct interaction with the aerospace industry, is outlined. The current status of the free-piston shock tunnel laboratory at the Graduate Aeronautical Laboratories (shock tunnel T5) at Caltech are briefly described.

1. Drivers of high-enthalpy gas dynamics research.

When an aerospace vehicle travels through the atmosphere at hypersonic speed U, the specific kinetic energy of the free stream relative to the vehicle is $U^2/2$. At the surface of the body the flow is brought all of its ordered kinetic energy is converted to disordered or thermal energy. If this thermal energy is sufficiently high, a significant fraction of the molecular collisions is of such high energy as to cause molecules to become vibrationally excited and to dissociate, possibly even to ionize. Thus, in the vicinity of the vehicle there exists a flow field in which processes of vibrational excitation and of dissociation (or other chemical processes) are important. They occur at rates which are peculiar to the gas properties and the local conditions. If the rates are very large compared to the rate of flow, the gas is in thermodynamic equilibrium everywhere. Otherwise the additional complication of thermodynamic nonequilibrium must be taken into account. Thus the behavior of the gas is vastly more complex than that of a perfect gas valid at lower speeds.

Such "real gas effects" are clearly of two kinds:

- Because $U^2/2$ is comparable to the specific dissociation energy D, dissociation occurs and the gas can no longer be treated as a perfect gas even if in thermodynamic equilibrium.
- If the rate of dissociation is finite, nonequilibrium effects introduce a time scale that is a property of the gas and the conditions.

To fix ideas, we focus on dissociation, though the same discussion applies throughout to other rate processes.

Those applications in which vehicles travel sufficiently fast to cause chemical reactions to occur in the gas include

- transport to and from space through the earth's atmosphere, orbital transfer
- entry into the atmospheres of the planets or some of their moons, space maneuvers involving aerodynamic braking
- very high speed missiles

Caltech takes the long term view appropriate to a university, that the first two items in the above list will continue to increase in importance over the next 20 years. The need for detailed knowledge of and well-educated specialists in the field is therefore the specific driver for the establishment of a high enthalpy laboratory, at a university such as Caltech.

2. Experimental methods of high-enthalpy gas dynamics.

2.1 Simple scaling laws of real-gas effects.

To focus ideas, it is convenient to consider as a specific problem of high-enthalpy gas dynamics that of the flow over a blunt body, suc radius L, in the vicinity of the front stagnation streamline, see Figure 1. In the case of transport through the earth's atmosphere a typical flow speed for this case is $U = 7km/s$. Figure 1 contrasts the case when the gas may be considered to be a perfect diatomic one, with that of the real gas. As may be seen, the temperature profile along the stagnation streamline exhibits very large real-gas

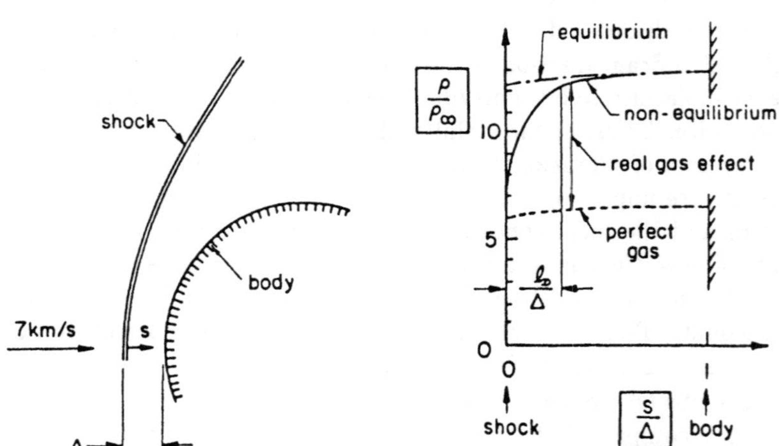

Figure 1 Density profile along the stagnation streamline in a perfect diatomic gas, an equilibrium flow and a nonequilibrium flow. Note the large increase in the density due to real-gas effects and the characteristic length l_D associated with nonequilibrium dissociation.

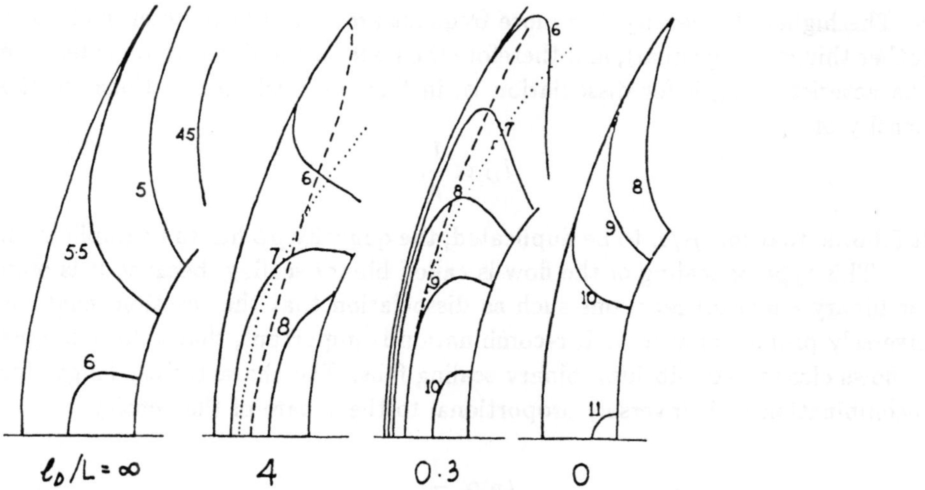

Figure 2 Density field in the shock layer of a circular cylinder in nitrogen flow at $6km/s$ for different values of the parameter l_D/L. ∞ corresponds to frozen flow (diatomic perfect gas) and 0 corresponds to equilibrium flow. 4 and 0.3 represent relatively slow and fast dissociation. The full lines show lines of constant density indicated in multiples of free-stream density. The dashed and dotted lines are representative temperature and composition contours (from Ref. 1).

effects related both to equilibrium and to nonequilibrium effects. The latter introduce the characteristic length of dissociation l_D related to the dissociation time scale by the flow speed.

To generate the same physical effects and flow structure in a test as occur in the real situation of a vehicle flying through the earth's atmosphere at $7km/s$, it is necessary that the dimensionless parameters $U^2/2D$ and l_D/L take the same values in the tests as in the real flow. A ground test facility, such as a wind tunnel, therefore needs to provide for the proper duplication specifically of these two, as well as other dimensionless parameters, in order to exhibit comparable real-gas effects. Figure 2 shows just how dramatic the effects of the parameter l_D/L can be in the density field in front of a circular cylinder computed at a specific value of $U^2/2D$, for l_D/L ranging from 0 (equilibrium flow) to ∞ (perfect gas).

The requirement that the flight value of $U^2/2D$ be duplicated in the test facility demands that U be duplicated, since D is a property of the gas, and the many features of the gas properties demand that the same gas be used. D is therefore not free to be chosen. Thus, a flow at a speed of $7km/s$ needs to be generated in the laboratory.

The requirement that the flight value l_D/L be duplicated is intimately linked with the need to test at a reduced geometrical scale. Thus, if the real vehicle is scaled down by a factor of 100 for laboratory tests, the dissociation length l_D also has to be scaled down by the same factor. The implications of this requirement become clear in the light of the dependence of l_D on the density

ρ. The higher the density, the more frequent are collisions between molecules (other things being equal), and therefore the faster is the dissociation rate. The characteristic length for dissociation is, in fact, inversely proportional to the density or

$$l_D \propto \frac{1}{\rho}.$$

It follows, that for l_D/L to be duplicated, the quantity ρL has to be duplicated.

This type of scaling of the flow is called binary scaling, because it is only for binary chemical reactions such as dissociation that the reaction length is inversely proportional to ρ. If recombination is important, that is for example in flows close to equilibrium, binary scaling fails. The characteristic length for recombination l_R is inversely proportional to the square of the density

$$l_R \propto \frac{1}{\rho^2}.$$

Fortunately, there exists a very large class of important flows in which recombination is so slow as to be irrelevant, and in which it introduces only small scale effects. Nevertheless, it must always be kept in mind that binary scaling is an approximation.

If the quantities U and ρL are both duplicated, the temperature and composition fields will be reproduced in the external flow field of a body. Hence the viscosity field μ, which depends essentially only on the temperature, will also be reproduced. Thus the Reynolds number

$$Re = \frac{\rho L U}{\mu},$$

which is another important dimensionless parameter, will automatically be reproduced. Exceptions to the applicability of binary scaling occur whenever recombination is important. Examples are boundary layer flow with a catalytic wall and external flows in the atmosphere of Venus.

Two further dimensionless parameters of some importance are the Mach number M and the free stream chemical composition. In some forms of experimental facilities the flight values of these two parameters may not be duplicated correctly. This introduces further problems as will be discussed in section 3.2.

To summarize: For a large class of flows involving high-enthalpy real-gas effects in earth atmospheric maneuvers binary scaling provides a method whereby the essential features of dissociating gas flow may be simulated in the laboratory provided that flight values of U and ρL are duplicated.

2.2 Rationale for simulation facility.

If an air flow at $7 km/s$ is brought to rest at a stagnation point, the temperature reached is approximately 8500K. Conversely, if such an air flow is generated by expanding it through a nozzle from a reservoir at rest, the reservoir temperature has to be approximately 10000K. These are such high temperatures that no material exists to contain the gas for a prolonged period, especially if the pressure in the reservoir is high. It is therefore essential to produce the reservoir conditions for a sufficiently short time to avoid melting and evaporation of the containing material. This limits the testing time, of course, and imposes constraints on the types of experiments that may be performed. The alternative of cooling the walls leads to serious deficiencies in the quality of the nozzle expansion.

A well proven method of heating and compressing a gas at the same time is to pass a shock wave over it and to reflect the shock from a solid surface to heat and compress the gas even more by a second passage. To produce a reservoir at conditions such that the nozzle expansion provides a speed of $7km/s$, the speed of the primary shock has to be $5km/s$.

A device of this type, where a shock reflection is used to provide the reservoir for a nozzle expansion to produce a high speed flow, is called a shock tunnel. It consists of a driver tube initially containing driver gas at high pressure and separated from a shock tube containing test gas at low pressure by a strong main diaphragm, and a nozzle and test section initially evacuated and separated from the shock tube by a weak secondary diaphragm. To operate the shock tunnel, the main diaphragm is broken, a shock wave travels through the shock tube, thus heating the test gas. The shock is reflected from the end of the shock tube, where it also breaks the secondary diaphragm. The doubly heated and compressed test gas is thus able to expand and accelerate through the nozzle to the appropriate speed and density. The performance of a simple shock tube, where driver and driven tube have the same cross sectional area may be written down explicitly in the form

$$\frac{p_4}{p_1} = f(M_s, \frac{a_4}{a_1}, \gamma_4, \gamma_1)$$

where p_4 is the driver gas pressure, p_1 is the test gas pressure, M_s is the shock Mach number, a_4 and a_1 are the driver and test gas speeds of sound and γ_4 and γ_1 are the driver and test gas ratios of specific heat. (This formula does not take into account the real gas effects occurring in the flow after the primary shock and the non-ideal operation of many facilities, but these do not modify the qualitative results of this argument).

In order to discuss the implications of the shock tube formula, it is best to show it in a diagram, see Figure 3. For a particular pair of values of γ_4 and γ_1, Figure 3 shows M_s plotted against p_4/p_1 with a_4/a_1 as a parameter. Since the shock speed has to be $\sim 5km/s$, this requires M_s to be ~ 15. Figure 3 shows a dashed line at $M_s = 15$, giving a target value to be reached or exceeded. This may clearly be done in two ways: By raising the pressure ratio p_4/p_1, or by raising the speed of sound ratio, or both. The pressure ratio is limited by two requirements. First, the maximum safe pressure is clearly a cost factor and also an operational factor. A reasonable maximum for safe operation in a university environment is 200 MPa. Second, the requirement for ρL duplication dictates increased density for reduced scale. The density is proportional to p_1, so p_1 must not be too small. A reasonable minimum is 10 kPa. Thus, an upper limit on p_4/p_1 is 20000. This is shown as a chain-dotted line in Figure 3. To achieve $M_s \geq 15$ at $p_4/p_1 \leq 20000$, Figure 3 shows that $a_4/a_1 \geq 10$.

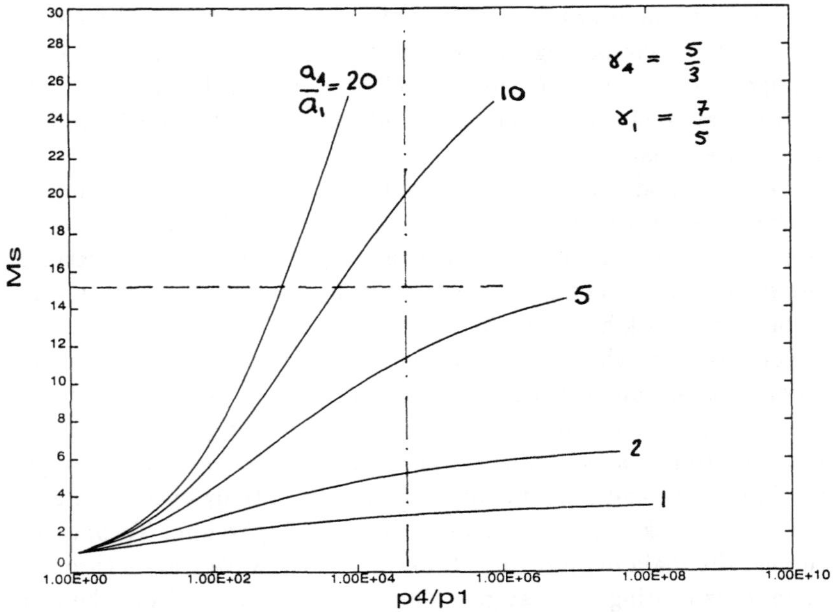

Figure 3 The shock tube formula (see Ref. 2). Safety and flow density restrictions require the pressure ratio p_4/p_1 to lie below 20,000 (chain-dotted line). Enthalpy requirements demand the shock Mach number M_s to reach or exceed 15 (dashed line). For helium driver gas ($\gamma_4 = 5/3$) and air test gas ($\gamma_1 = 7/5$) this dictates that a_4/a_1 be 10 or greater.

It is therefore necessary to produce a high speed of sound in the driver gas. The speed of sound a_4 is related to the temperature T_4 and molecular weight $\mathbf{M}_4$ by

$$a_4^2 = \gamma_4 R T_4 / \mathbf{M}_4,$$

where R is the universal gas constant. A light (low $\mathbf{M}_4$) and hot (high T_4) gas is called for. To produce $a_4 \geq 3400m/s$ ($a_4/a_1 \geq= 10$), in helium ($\mathbf{M}_4 = 4$) requires $T_4 \geq 3500$K. This is also a temperature considerably in excess of the boiling point of stainless steel, so that even the driver gas has to be heated in a transient way.

A clean and repeatable way of heating a gas is to compress it adiabatically by a heavy piston which may be accelerated in a tube, for example by compressed air. If a helium driver gas is heated by a piston compression, a temperature of 4500 K may be achieved with a volume reduction by a factor of 60. This simultaneously raises the pressure by a factor of 1000, so that, to achieve a driver pressure of 150MPa an initial helium pressure, before piston compression, of 150kPa is enough. Clearly, the compression tube has to be long so that the remaining driver gas length at the instant of diaphragm burst is not too small. Also, a number of other requirements including safe piston deceleration complicate the processes occurring during the generation of the shock wave.

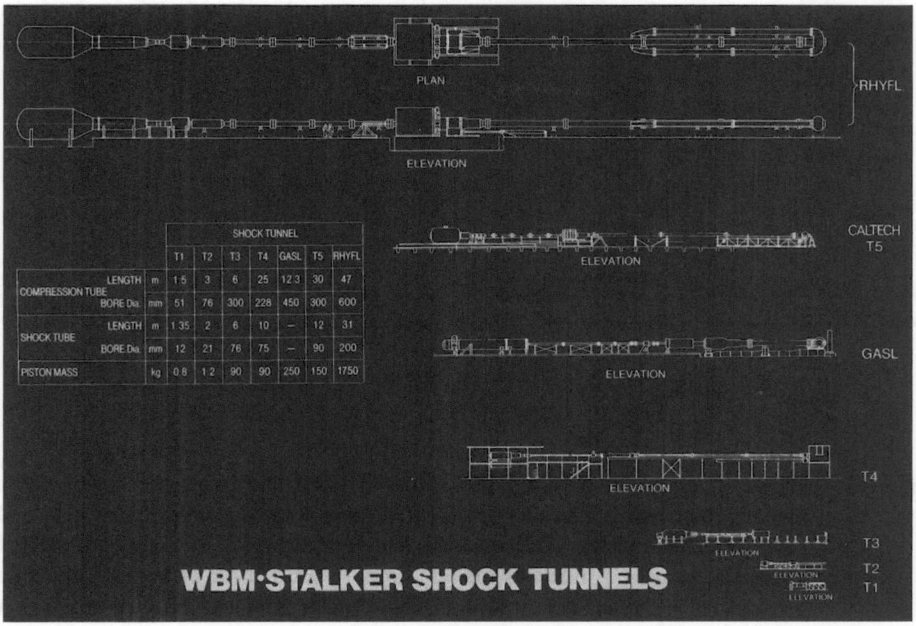

Figure 4 Scale increases in some free-piston shock tunnels. The approimate completion dates of these facilities are

T1	T2	T3	T4	T5	GASL	RHYFL
1962	1964	1969	1987	1990	1991	1991

From Ref. 3

A number of shock tunnels using the free-piston driver technique have been built, and several are now under construction. The existing ones known as T1, T2, T3 in Canberra, Australia and T4 at Brisbane, Australia were developed by Stalker from 1961 to 1987. They involved a sequence of scale increases, each of which helped understand the problems associated with size and to find solutions. An important lesson that was learned from this development was that it pays to go through this process of successive upscaling, because even the small steps taken caused many problems. Figure 4 is a brochure of the Australian company WBM-Stalker showing the four existing shock tunnels and three of the facilities presently under construction drawn to the same scale, together with their parameters. A further facility being built by the DLR at Gö ttingen, W. Germany, known as HEG, lies between T5 (the shock tunnel being built at Caltech) and RHYFL in scale (refer to Figure 4).

Many of the important discoveries of high-enthalpy real-gas dynamics were made during the 1970's in the facility T3. A large number of interferometric flow measurements made in that period still provide the best data for computer code calibration at high enthalpy (see section 4).

It is important to discuss briefly alternative methods of producing a fast gas flow with a shock tube as generator. Instead of using a shock reflection

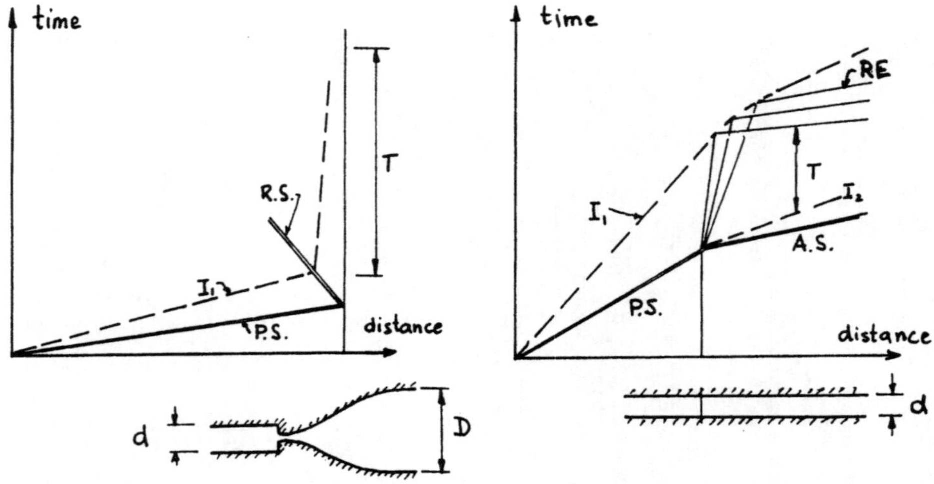

Figure 5 Wave diagrams of reflected shock tunnel (left) and expansion tube (right). Double lines indicate shocks: P. S., R. S. and A. S. denote primary, reflected and accelerated shock. Dashed lines denote interfaces. R. E. is the reflected expansion wave. T indicates the test time. Note that the time scale has been expanded for the expansion tube to show the features better. The expansion tube has a shorter test time and a smaller flow diameter, but gives lower free stream dissociation and more rapid flow start.

to stop the test gas, after heating and accelerating it with the primary shock, and then accelerating it again with a steady nozzle expansion, this gas may be accelerated further by using an unsteady expansion. A device of this type is called an expansion tube. The two principles of operation are compared in corresponding wave diagrams in Figure 5. Without going into detail, the advantages of the expansion tube principle are, that it provides a more accurate reproduction of the free stream composition of the flow (a nozzle expansion is usually not able to fully recombine the gas dissociated by the reflected shock) and that it provides a much more sudden start to the flow (an advantage for heat transfer measurements). One pays for these advantages by smaller size of flow and shorter test time. The facility under construction at GASL (see Figure 4) is a free-piston expansion tube. Without excessive modifications, T5 may be converted to expansion tube operation.

Piston facilities depart from ideal shock tube operation, because the effective driver tube is short. This is compensated for to a large part by the fact that the piston continues to move after diaphragm rupture and by the fact that the driver tube area is about 10 times that of the shock tube.

A problem in shock tunnel operation is the termination of the test time through driver gas contamination. The onset of contamination depends critically on the interaction of the reflected shock with the shock tube boundary layer and with the driver/test gas interface.

In order to characterize the performance of different facilities in the context of the high-enthalpy real-gas effects, it is best to show their limits on a plot of ρL

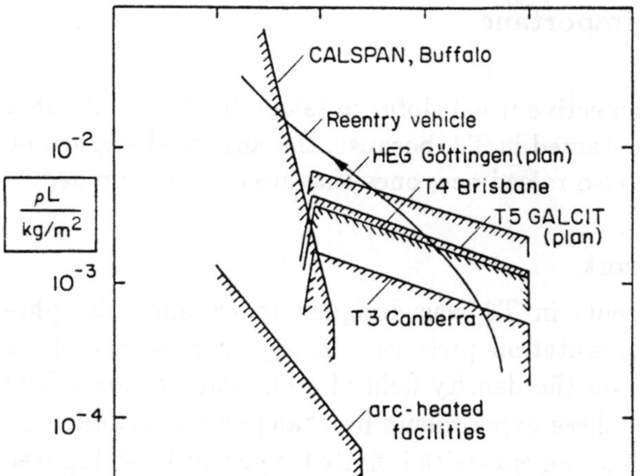

Figure 6 Performance envelope of some facilities for binary scaling of real-gas effects in air, compared with an earth entry trajectory. (From Ref. 4)

against U. This is done in Figure 6 for the free-piston facilities. Development work currently in progress at other laboratories aims to increase the pressure of arc-heated facilities which is presently limited to 20MPa. The projected facility F4 of the ONERA at Toulouse, France aims at a reservoir pressure of 200MPa. So far, the free-piston facilities are the only ones to have achieved the conditions of the critical part of the reentry trajectory which also approximates other aerospace vehicle operations (see Figure 6).

2.3 Flight testing

Clearly, a number of problems of importance may not be studied in a shock tunnel. The short running time eliminates problems involving low frequency geometric adjustments such as are important in engine control. It even eliminates problems associated with large separated flow regions. The large level of disturbances associated with the method of flow generation makes the investigation of transition to turbulence questionable (though both laminar and turbulent flows are reproduced correctly) because transition depends very critically on the level and spectrum of disturbances.

For these reasons it is necessary to complement ground-based experiments with flight tests. Because of their extremely high relative cost it is also necessary, however, to prepare flight tests by extensive ground-based experiments to first eliminate those problems that may be answered cheaply. The approximate ratio of cost per launch for a free flight test to cost per shot in T5 for a model of the same size is 1000, without instrumentation. Since the instrumentation is much more expensive for flight tests, this number is almost doubled.

3. Particular problems of importance

To put this topic into perspective it is helpful to take a brief look at some of the results that have been obtained in T3, because, tho answered important scientific questions, they have also raised new ones that need to be pursued.

3.1 Selected results from T3 work

One of the first experiments in T3 was designed to examine the phenomenon exhibited by the computation presented in Figure 2, of the effect of dissociative nonequilibrium on the density field of a circular cylinder. The parameter l_D/L was varied in these experiments by changing the cylinder diameter $2L$. Interferometric measurements with infinite fringe width setting (see Figure 7) showed that the dramatic departure of the fringe shapes (equivalent to lines of constant density) from those of perfect gas or equilibrium flow is indeed observed. The change in l_D/L also caused the expected changes. Interesting additional features of these interferograms are that the density rise in the laminar boundary layer is clearly evident and that, in the case of the 2 inch diameter cylinder a separation line is clearly visible at the rear. A viscous flow simulation was recently performed to compare with these cases and is presented in section 4. These were the first detailed results of strongly dissociating blunt body flows ever measured.

The understanding of these flows gained through the experiments led to a theoretical investigation of the strongly dissociating region near a curved shock wave. This was based on the asymptotic treatment of the case when $l_D/L \rightarrow 0$, i.e., when the thickness of the relaxing layer is small compared with the curvature radius of the shock. This treatment is similar to a boundary layer approach. The results of this study, see Figure 8, led to the very important conclusion that, for a wide range of conditions, the dissociative reaction is stopped because of the cooling resulting from the shock curvature. This effect was called dissociation quenching. A consequence of this effect is that, after the thin reacting layer, the temperature is almost independent of the shock angle β and the density varies as $sin^2\beta$. This is in contrast to the case of perfect gas flows, where the opposite is the case. The results of this work have very significant value for the designer who can estimate the qualitative effect far downstream of the dissociative flow near the nose of a blunt slender body, in a region where he might wish to place a control surface.

A further simple flow is that over a wedge, in which the effects of chemical nonequilibrium can be very dramatic. An example of the simple case of flow with an attached shock is shown in Figure 9 in the form of a finite fringe interferogram. The fringe displacement normal to the fringe in the free stream is a measure of the density change. As may be seen, a strong dissociative nonequilibrium effect is evident in the density rise after the shock wave, that occurs over a finite distance l_D before the fringes (and the density) level out again near the body. Again the density rise of the laminar boundary layer may be seen very near the body.

Figure 7

Interferograms of nonequilibrium nitrogen flow at 5.6 km/s over circular cylinders 1, 2 and 4 inches in diameter obtained in T3. The infinite fringe width setting yields fringes that approximate lines of constant density. Comparing these with figure 2, it is clear that strong nonequilibrium dissociation is present and the expected changes due to the variation of l_D/L over a factor of 8 occur. Note the fringe shift due to the density rise in the boundary layer and the separation line in the middle photo. (From Ref. 1)

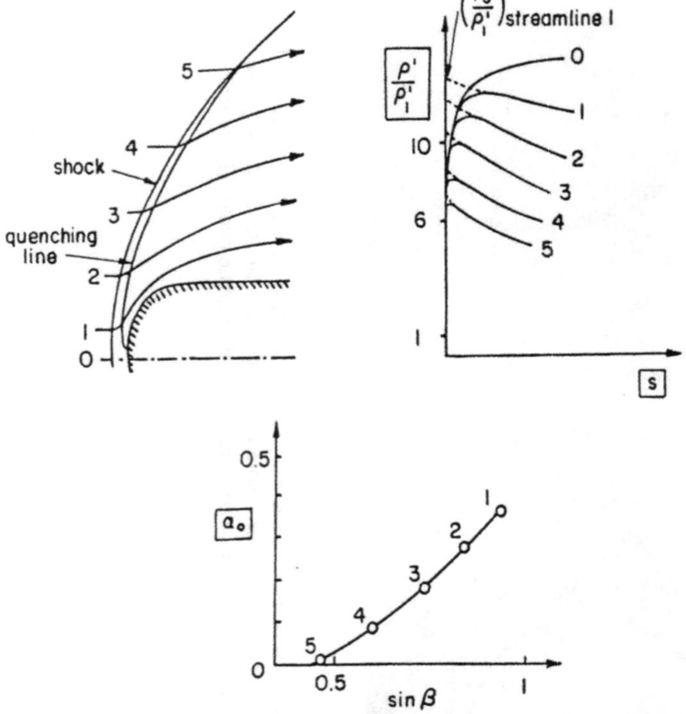

Figure 8 The dissociation quenching effect. Along 5 selected streamlines (top left) that pass through the curved shock and dissociation region before quenching, the top right shows the density profile. The quenched dissociation fraction varies with the shock angle at which the streamline traversed the shock as indicated (bottom). (Ref. 4 and 5)

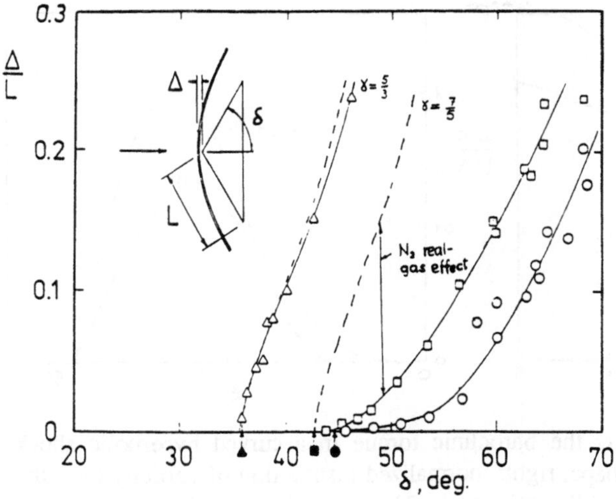

Figure 10 Shock detachment from a wedge. Δ - perfect gas argon; □ - nitrogen, 5.6 km/s; ○ - carbon dioxide, 5.1 km/s. Full line: faired through experimental points dashed line: approximate theory for frozen flow. Filled symbols: theoretical detachment angle. (From Ref. 7)

On the basis of a theoretical argument it was predicted in 1977 that the shock detachment distance on a symmetrical wedge would behave very differently in a nonequilibrium flow than in a perfect gas flow. In particular, the detachment distance was predicted to increase very much more gradually from zero. This was indeed observed in both nitrogen and carbon dioxide flows as is shown in Figure 10. The dramatic nonequilibrium real-gas effect exhibited by these results has profound significance for practical vehicle design. Recent computations to compare with these data are presented in section 4.

3.2 Important future experiments

One of the special features of hypersonic flows is that the curved shock on the blunt nose of a vehicle can generate such high levels of vorticity, that an unstable shear layer develops, which may have dramatic effects on the stability of the boundary layer far downstream. The vorticity on the stagnation streamline and on a streamline far outboard are both zero, and a maximum occurs on a streamline somewhere between them. This is illustrated in Figure 11 showing the vorticity generated as a function of the shock angle for a hyperbolic shock shape in a perfect gas. As may be seen, the maximum vorticity increases rapidly with Mach number. In fact, at around $M = 10$ the maximum is roughly comparable with the vorticity in the boundary layer, depending (slightly) on the Reynolds number. A very important real-gas effect on this flow is that

Figure 9 Finite-fringe interferogram of carbon dioxide flow over a wedge at 4.1 km/s. Note rise of fringe shift through the dissociation layer near the shock with subsequent equilibration and rise through the thin boundary layer. (From Ref.6)

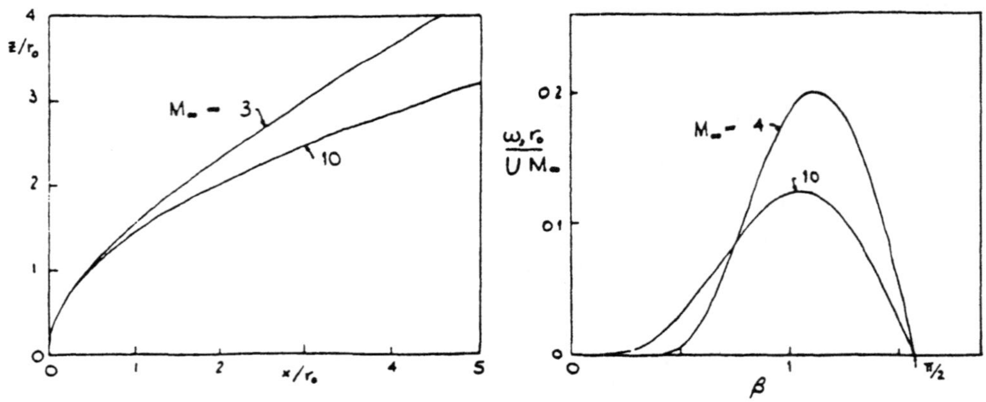

Figure 11 Vorticity produced by the baroclinic torque in a curved hyperbolic shock wave. Left: shock shape, right: normalized distribution of vorticity over the shock angle. (Perfect diatomic gas). Note that the vorticity ω_2 has been divided by the free stream Mach number to be able to plot the two curves on the same diagram. The vorticity maximum increases rapidly with M_∞.

the density ratio across the thin reacting layer near the shock is dramatically increased by them ($\approx$ factor of 2). Since the vorticity generated at a curved shock is approximately proportional to the density ratio, it may be expected that the shear layer will have roughly twice the vorticity in a real gas, and the phenomenon gains in importance. This is clearly a problem that needs to be examined. It is also known under the name entropy layer swallowing, since the shear layer generated represents the outer edge of a region of high entropy, which, far downstream, is eventually engulfed by the boundary layer.

A second important problem is the effect of more complex gas composition (than in the case of nitrogen) on the phenomenon of dissociation quenching after a curved shock. In particular, the intermediate gas NO has a profound effect in the case of the chemistry of air and may cause the relative importance of recombination to increase because of an alternative two-stage path. Such an investigation, like many that are planned, requires intimate coupling of the experiments with a massive computational effort, in which parameters are varied over significant ranges to produce an understanding of the essential and to remove extraneous effects.

Very especially, the higher pressure of T5 vis-a-vis T3 will almost certainly put us in the position of being able to produce a fully developed turbulent boundary layer in high enthalpy flow. We propose to examine the structure of the layer and the energy budget both by optical techniques that integrate along the line of sight (such as interferometry) and by inclined planar laser-induced fluorescence (PLIF) as well as by wall and profile measurements. Through parameter variations over wide ranges it is expected that qualitative real-gas effects will be discovered and it may be possible to order and classify these. Clearly, this must be an experiment of the exploratory kind that can lead to a well-designed specific experiment only in a second stage.

A very important practical problem of NASP and other vehicles is that of the heat load on the structure. A significant fraction of the effort of the T5 research team must therefore go into the development of satisfactory methods of heat transfer measurement. In a second effort in this direction, the method of thermal protection known as film cooling poses extensive uncertainties and must also be studied in the real-gas environment. Here too, the diagnostic technique PLIF (possibly with seeding of the cooling film) is intended to be applied.

A fifth problem of immediate interest to Rocketdyne for the air-breathing and supersonic combustion engine of NASP is that of turbulent mixing and combustion. To make a fundamental study of this, it is planned to inject hydrogen into the flow with a jet synchronized with the main flow. The technique of achieving this has been determined and partially designed. It involves a second, smaller shock tube to generate the hydrogen jet.

Parallel to these investigations, improvements of the flow quality and duration in T5 should be investigated. Minor reconfigurations of the detail near the ends of the shock tube may be expected to lead to significant improvements and need to be tested.

4. Philosophy of the approach

It is clear that complete simulation in the laboratory of the flow on an aerospace vehicle is not possible. It is also clear, however, that very important discoveries can be made by studying flows that exhibit the main features of the real-gas effects in a fundamental approach. This must be done by getting as close as possible to the real flow by using the proper scaling laws.

The second feature of the approach must be to couple with the experiments a massive computational effort, so that the different computer codes may be calibrated against well-defined experiments with established and specified boundary conditions. Only in this manner can confident extrapolation be made to any real flows.

In both experiments and computations a commonly encountered failing is that no significant parameter variations are made. In the field of real-gas experimentation this parameter variation is of particular importance since the number of possible extraneous effects that need to be eliminated is much greater than in flows in more conventional regimes.

Whenever possible, it is necessary to make measurements carrying not only point information but areal information. For the purpose of resolving unsteady effects such as turbulent structure in shear or boundary layers this is essential. Further, traditional methods, such as schlieren photography or interferometry, which average the information over the line of sight, have fortunately been complemented by techniques such PLIF that resolve spatially along the line of sight. These are eminently more effective methods for comparison with computation.

Figure 12 Interferograms of the transition from regular to Mach reflection of shock waves. two wedges are arranged symmetrically to the flow so that the shocks generated reflect from the symmetry plane. On the left, they reflect in the regular and on the right in the Mach reflection mode. Top: nitrogen, bottom: carbon dioxide. In both cases the difference between the wedge angles left to right is less than 0.2 degree. (From Ref. 8)

Finally, it cannot be emphasized too strongly that sensitive indicators of the effects studied must be the ones to measure. To illustrate this principle, consider the blunt body flows of Figure 2 again. A dramatic effect is exhibited by the density field as a result of l_D/L variation. If, instead, the pressure field had been plotted, it would exhibit only very minor quantitative changes that would certainly be only within the noise of the measurement errors. Hence, density is clearly the variable to measure in this case. In another example shown in Figure 12, the shock configuration can be a sensitive indicator.

To illustrate first efforts to couple modern computation with this work, Figure 13 shows comparisons of two of the flows of Figure 7 with computations by Candler. More recently, Candler at our suggestion also computed the shock detachment from a wedge with spectacular results, see Figure 14. It is very important to note that Candler discovered faults and errors in his grid and code only through comparison with experiment, so that the need for the combined approach is evident.

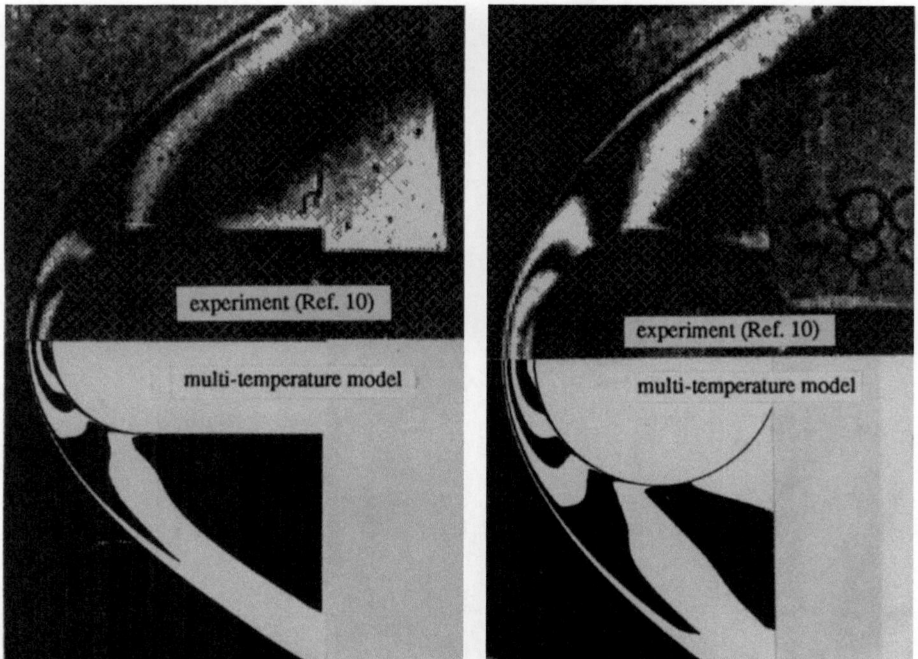

Figure 13 Comparison of experimental and computed flow of nitrogen over a circular cylinder (see also Figure 7) by Candler. In each of the two cases the top half is the experimental interferogram (Figure 7) and the bottom is the corresponding computed one. All the special features of the flow are faithfully reproduced by the computation. (From Ref. 9)

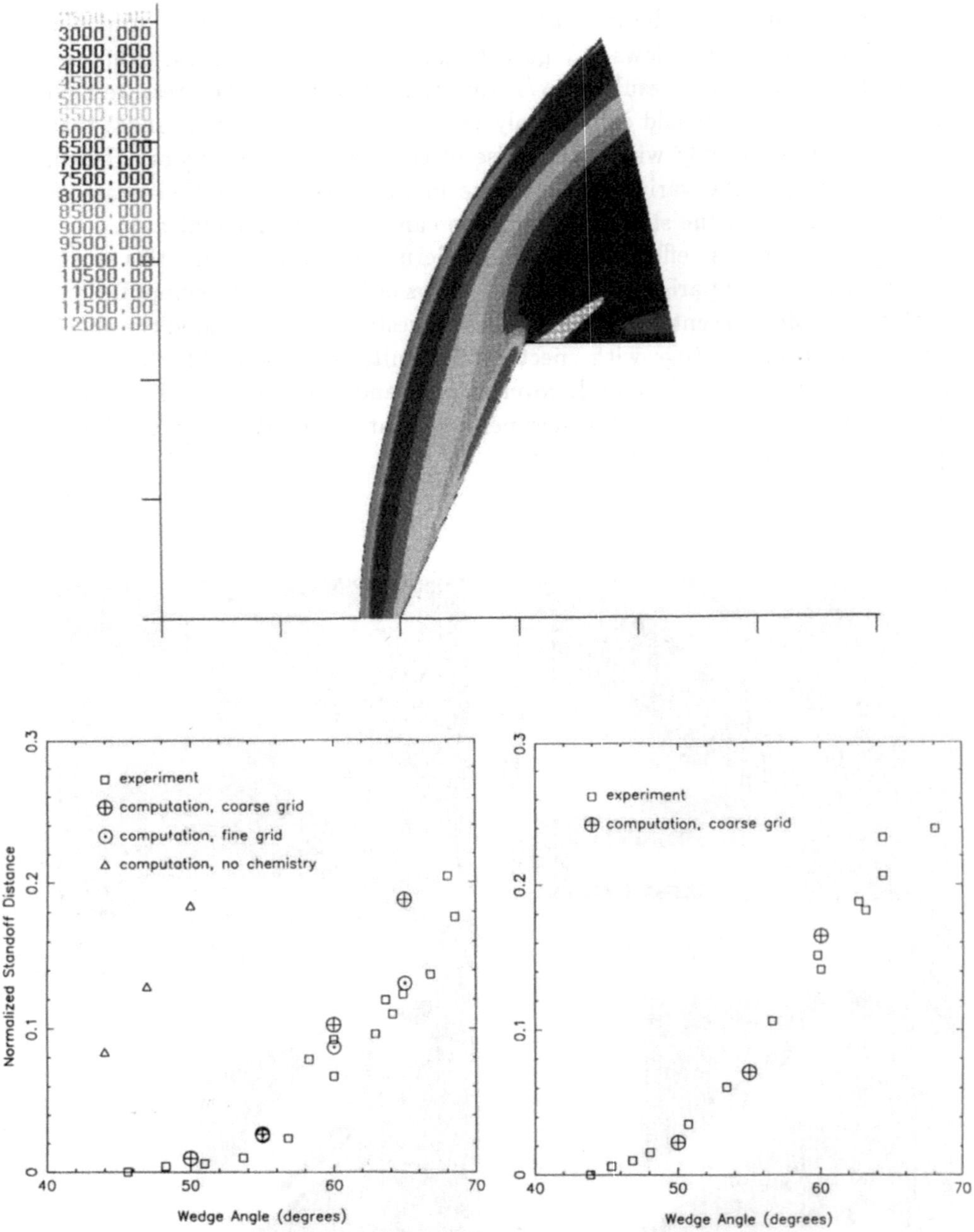

Figure 14 Candler's computation of the flows shown in Figure 10.

a. Temperature field in CO_2 flow over a 65 deg helf-angle wedge (fine grid, 110 x 100). Note high temperatures near the shock and in the boundary layer.

b. Carbon dioxide. Note frozen flow points.

c. Comparison of stand-off distance experiment and computation: nitrogen.

(From private communication by Candler)

Figure 15 Photograph of the T5 laboratory.

5. Status of the T5 lab

The construction of the new laboratory building on the roof of Guggenheim building including utilities is complete. Part of the roof of the new building has been left open to allow the hardware to be brought up by crane. The roof location has been chosen because this was the only place at GALCIT where a straight 200 ft length could be found. Figure 15 shows a photograph of the laboratory.

The design of the facility is complete, several parts have been manufactured and are in storage, the remainder are being fabricated by different companies. Figure 16 shows an overall drawing of the facility and a representative photograph of a part is presented in Figure 17.

The data acquisition system has been selected. Orders are about to be placed. The system provides for 30 channels of digital recording in addition to the channels necessary for the operation and run documentation.

A bare-bones start to an optical diagnostics system consisting of schlieren and interferometry capability has been designed, and orders are about to be placed. Financial constraints presently prohibit the development of a PLIF system. However, discussions with Rocketdyne indicate that it may be possible for us to use the system developed by them for some of our fundamental research projects during a limited period.

None of the gas dynamical calculations necessary for the operation of a high-enthalpy shock tunnel can be done without making fairly sophisticated

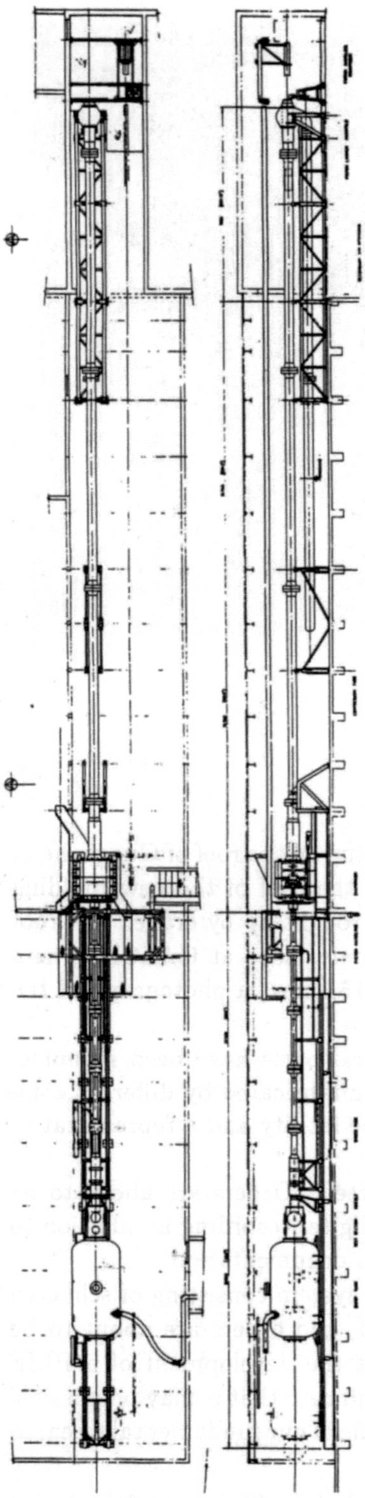

Figure 16 Overall drawing of T5 shock tunnel.

Figure 17 Photograph of launch manifold for T5. This is the part at the extreme right of Figure 16.

statistical mechanical computations. To this end, a number of computer codes have been written and tested against existing simpler programs. In particular, a program for inviscid nozzle flow computations in two dimensions and for finite rate chemistry is now operational. This is sufficiently fast and economical to be built into a design loop for optimizing nozzle shape. An example of partial output of this program for the T5 axisymmetric nozzle is presented in Figure 18.

6. References

1. Hornung, H. G. Non-equilibrium nitrogen flow over spheres and circular cylinders. *J. Fluid Mech.* 1972, 53, 149-176.
2. Liepmann, H. W. & Roshko, A. *Elements of Gasdynamics.* Wiley, 1957.
3. Morrison, R. To appear in Proceedings, 17th International Symposium on Shock Tubes and Waves, Lehigh, 1989.
4. Hornung, H. G. 28th Lanchester Memorial Lecture - Experimental real-gas hypersonics. *Aero. J.,* December, 1988.
5. Hornung, H. G. Nonequilibrium dissociating flow behind a curved shock wave *J. Fluid Mech.* 1976, 74, 143-160.
6. Ebrahim, N. A. & Hornung, H. G. High-enthalpy non-equilibrium carbon dioxide nozzle and wedge flows *AIAA J.* 1975, 13, 845-846.

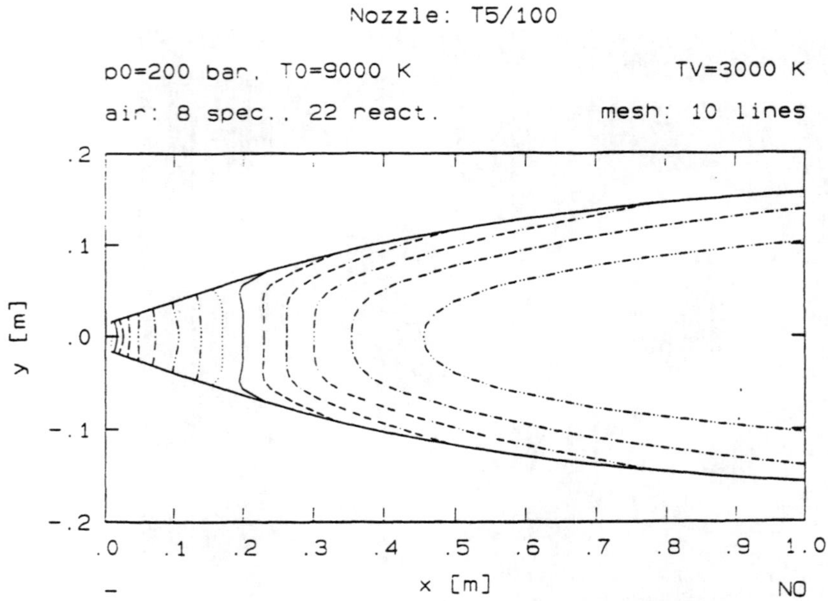

Figure 18 Example of computation of the T5 nozzle (NO concentration contours). (Ref. 10)

7. Hornung, H. G.& Smith, G. H. The influence of relaxation on shock detachment. *J. Fluid Mech.* 1979, 93, 225-239.
8. Hornung, H. G., Oertel, H. & Sandeman, R. J. Transition to Mach reflexion of shock waves in steady and pseudo-steady flow with and without relaxation. *J. Fluid Mech.* 1979, 90, 541-560.
9. Candler, G. On the Computation of Shock Shapes in Nonequilibrium Hypersonic Flows. 27th Aerospace Sciences Meeting, Reno, NV, AIAA-89-0312, 1989.
10. Rein, M. Surf: A Program for Calculating Inviscid Supersonic Reacting Flows in Nozzles. GALCIT FM 89-1, California Institute of Technology, 1989.

PROBLEMS FOR ANALYSIS

In the perspective of European and American space projects, the knowledge of flows at high Mach numbers and large angles of attack, occurring during the atmospheric reentry of space vehicles, has regained recently considerable interest. The extreme physical conditions realized during this critical phase are at the limit of possible experimental measurements, because of the rarefied atmosphere and/or the very high temperatures. The experimentation is therefore very difficult (if available) and costly. In addition, safety requirements cannot be achieved by over-dimensioning the structures without compromising the operational performance. Hence, the numerical simulation is led to play a major role in the process of the project validation.

During reentry, three main types of flow are encountered:

- rarefied gas flow,
- reacting flow,
- non-reacting flow.

The governing equations to be solved are then respectively - the Boltzmann equations, - the Euler or Navier-Stokes equations combined with a chemistry model, - the Navier-Stokes equations possibly combined with a turbulence model.

The following eight critical problems have been proposed to the participants:

Problem 1: Flow over a Slender Cone

Problem 2: Base Flow

Problem 3: Flow over a 2D Ramp

Problem 4: Flow over a 3-D Obstacle

Problem 5: Corner Flow

Problem 6: Flow over a Double Ellipse/ Double Ellipsoid

Problem 7: Flow over a Delta Wing

Problem 8: Non-equilibrium Flow in an Arc Jet or a Shock Tube

Due to the interest or the difficulty of each test case, a level of gradation, ranging from no "star"(0) to four "star (****) has been attibuted. The mandatory condition for participation is to collect a sum of at least four stars, by any combination of test cases solved (for example four cases with one star (*), or two cases with two stars (**), etc...).

Fluid Characteristics and Notations

- **Fluid characteristics**

1. **Non-reactive flows**

• The fluid is a perfect gas with specific heats c_p, c_v with ratio

$$\gamma = \frac{c_p}{c_v} = 1.4$$

• If p is the static pressure, the pitot pressure p_t is defined by :

$$\frac{p_t}{p} = \left(\frac{(\gamma+1)M^2}{2}\right)^{\gamma/(\gamma-1)} \left(\frac{\gamma+1}{2\gamma M^2 - (\gamma-1)}\right)^{1/(\gamma-1)}$$

• The viscosity coefficients λ, μ verify the Stokes relation:

$$3\lambda + 2\mu = 0$$

• The Prandtl number is constant: $Pr = \mu c_p/\kappa = .72$ where κ is the thermal conductivity coefficient.
• The viscosity coefficient μ is a function of temperature T in Kelvin according to Sutherland's law:

– If $T_\infty \geq 120K$:

$$\mu(T) = \mu_\infty \left(\frac{T}{T_\infty}\right)^{1.5} \left(\frac{T_\infty + 110.}{T + 110.}\right) \quad \text{if } T \geq 120K$$
$$\mu(T) = \mu(120)\frac{T}{120} \quad \text{if } T \leq 120K$$

– If $T_\infty \leq 120K$:

$$\mu(T) = \mu_\infty \frac{T}{T_\infty} \quad \text{if } T \leq 120K$$
$$\mu(T) = \mu(120)\left(\frac{T}{120}\right)^{1.5}\left(\frac{120. + 110.}{T + 110.}\right) \quad \text{if } T \geq 120K$$

μ_∞ is obtained from the Reynolds number : $Re_\infty/m = u_\infty \rho_\infty/\mu_\infty$, a value of temperature in Kelvin is required to adjust the constant in Sutherland's law (T_∞ in Kelvin).
• When a turbulent simulation is required, the choice of the model is completely free.

2. **Non-equilibrium reactive flows**

• A computation with Park's model is required. This model is described in Appendix 1 where the unit for the rate constants is $cm^3/mole/s$. In addition, participants are encouraged to also use more accurate physical models to assess the influence of the model.
• The body surface is supposed to be fully non-catalytic in all test cases.

• Contributors should be aware that Sutherland's law is not physically relevent and alternate models should be used.

3. **Equilibrium reactive flows**

• Similar models as for non-equilibrium reactive flows but with the assumption of local equilibrium.

4. **Rarefied gases**

• The internal energy model to be considered is the Larsen - Borgnakke model. Calculations should be performed for a monoatomic gas for comparison or if the participants have not implemented the Larsen - Borgnakke model.
• The fluid is characterized by the mean free path at infinity λ_∞.

• **Notations**
The following non dimensional coefficients are defined:
• Pressure coefficient: $C_p = \frac{p - p_\infty}{0.5 \rho_\infty u_\infty^2}$ where ρ_∞ and u_∞ are the density and the velocity at infinity.
• Skin friction coefficient: $C_f = \frac{\tau_w}{0.5 \rho_\infty u_\infty^2}$ (τ_w: wall shear stress)
• Heat flux coefficient: $C_h = \frac{q}{0.5 \rho_\infty u_\infty^3}$ where q is the heat flux; in the continuum regime: $q = -\kappa \nabla T \cdot n$ where n is the outward normal to the wall.
• Stanton number : $St = \frac{q}{\rho_\infty \, u_\infty \, c_{p\infty}(T_{0_\infty} - T_w)}$ where $c_{p\infty}$ is the free stream specific heat at constant pressure, T_{0_∞} is the reservoir isentropic stagnation temperature and T_w is the wall temperature.
• All dimensions are given in the meter, kilogram, second (MKS) system.

1 Problem 1: Flow over a Slender Cone

1.1 Description of problem 1

Geometry:

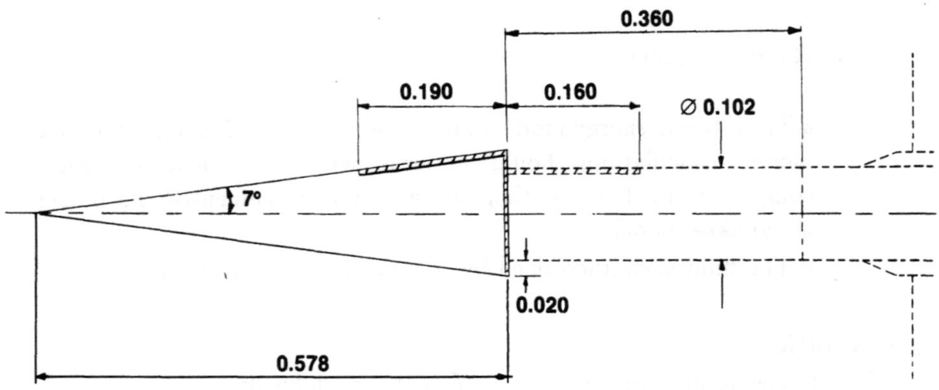

Flow measurement location

Model dimensions: semi-vertex angle: 7 degrees, base diameter: .142 m, sting diameter: .102 m, step height: .020 m, sting length .360 m.

Boundary conditions:

The following boundary conditions are prescribed :

- on the body, $u_{wall} = 0$ (adherence condition) ; the temperature on the wall T_{wall} is prescribed and constant.
- at infinity : Dirichlet boundary conditions are set according to the Mach number at infinity M_∞.

Test case:

	M_∞	Re_∞/m	T_∞	T_{wall}	Modelization	Level
1.1	9.16	$5.5\ 10^7$	59.8 K	290 K	laminar-transitional-turbulent	***
1.2	8.85	$1.53\ 10^7$	59.8 K	290 K	-	***

As forebody flow measurements, it is proposed to provide profile data from the forebody flow field at distances of .040 m and .180 m upstream of the base.

Heat transfer measurements are made along the cone surface, as far upstream of the base as possible.

1.2 Output formats

- **Wall values**
  - Cone side length, x axis, $\Delta L = 0.1$ represented by $2.5cm$ for $0 < L < 0.583m$
  - Static pressure on the y axis, $\frac{\Delta p}{p_\infty} = 1$ represented by $4.cm$
  - Stanton number on the y axis, $\Delta\ St = 10^{-3}$ represented by $10.cm$
- **profile data**
  - Normal displacement on the y axis, $\Delta y = 1.\ mm$ for $0 < y < 10.\ mm$ represented by $1.\ cm$.
  - Pitot pressure on the x axis, $\frac{\Delta p_t}{p_w} = 10.$ represented by $2.\ cm$.
  - Mach number on the x axis, $\Delta M = 1.$ represented by $2.\ cm$
  - Velocity on the x axis, $\frac{\Delta u}{u_\infty} = 1.$ represented by $10.cm$.
- **Contour plots**
  - Density ratio, scale 1/4, $\Delta(\rho/\rho_\infty) = 0.05$
  - Pressure ratio, scale 1/4, $\Delta(p/p_\infty) = 0.05$

2 Problem 2 : Turbulent Base Flow

2.1 Description of problem 2

Geometry

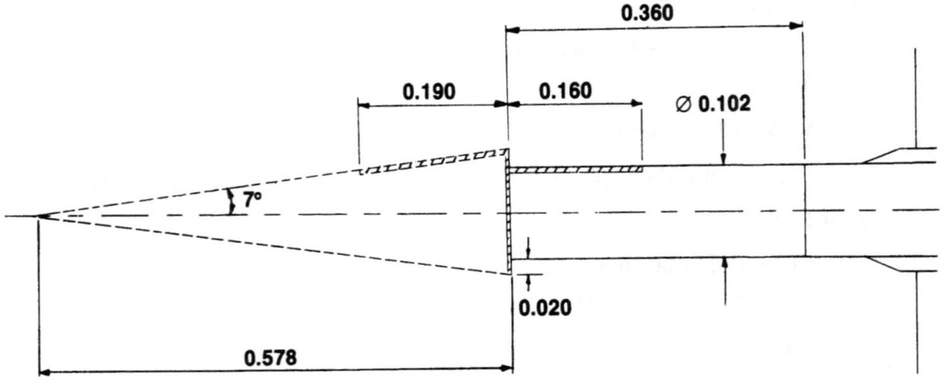

Boundary conditions :

- on the body: same as in problem 1.
- The inflow is prescribed using profiles obtained from experimental results or by solving problem 1.

Test case:

	M_∞	Re_∞/m	T_∞	T_{wall}	Modelization	Level
2.1	9.16	5.5 10^7	59.8 K	290 K	Turbulent-non-reactive	****

Pitot pressure profiles in the base flow region have been measured at three locations downstream of the base ($x = 50.\ mm$, $100.\ mm$ and $150.\ mm$).

2.2 Output Formats

Nomenclature - subscripts

- e condition at the edge of the boundary layer
- w conditions mesuared locally at the wall
- s conditions at the cone at separation
- b condition in the base region

- o stagnation conditions
- ∞ conditions in the free stream
- **Wall values**
  - distance from base on the x axis, $\Delta x = 0.1\,m$ for $-0.02\,m < x < 0.15$, represented by 10. cm. (x is measured w. r. t. base/sting corner)
  - static pressure on the y axis, $\Delta p_b/p_\infty = 1$ represented by 10 cm ; $\Delta p_b/p_s = 0.1$ represented by 2.5 cm
- **Profile data**
  - normal displacement on the y axis, $\Delta y = 5.\ mm$ for $0 < y < 50.\ mm$, represented by 1. cm.
  - pitot pressure on the x axis, $\Delta p_t/p_\infty = 10$ and $\Delta p_t/p_w = 10$ represented by 0.5 cm.
  - Mach number on the x axis, $\Delta M = 1$ represented by 1. cm.
  - velocity on the x axis, $\Delta u/u_\infty = 1$ represented by 10. cm
- **Contour plots**
  - density ratio, scale 1/2 from $(\rho/\rho_\infty)_{min}$ to $(\rho/\rho_\infty)_{max}$ using more than 40 contours.
  - pressure ratio, scale 1/2, from $(p/p_\infty)_{min}$ to $(p/p_\infty)_{max}$ using more than 40 contours.

3 Problem 3: Flow over a 2D Ramp

3.1 Description of problem 3

3.1.1 Geometry 1:

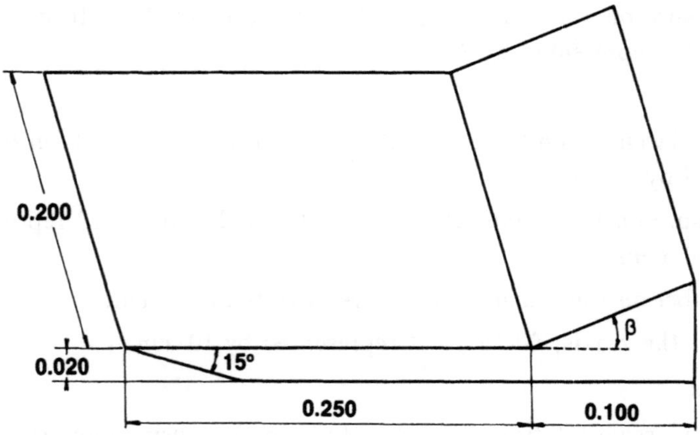

Boundary conditions: same as in Problem 1

Test cases:

	M_∞	Re_∞/m	T_∞	T_{wall}	β	Modelization	Level
3.1	5.	6. 10^6	80. K	288.K	0	Lam. n. react.	*
3.2	5.	6. 10^6	80. K	288. K	15	Lam. n. react	**
3.3	5.	4. 10^7	83. K	288. K	35	Turb. n. react	**
3.4	10.	3. 10^6	50. K	525. K	0	Lam. n. react	*
3.5	10.	9. 10^6	50. K	288. K	15	Lam. n. react	*

(Lam. = Laminar, n. react. = non reactive, Turb. = Turbulent)

3.1.2 Geometry 2:

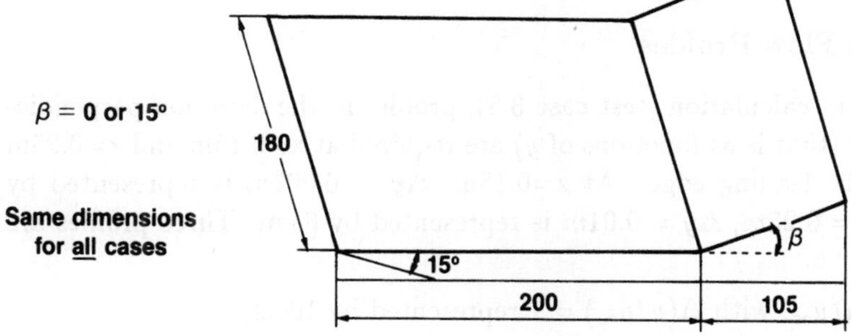

Test case:

	M_∞	Re_∞/m	T_∞	T_{wall}	β	Modelization	Level
3.6	15.	21. 10^6	48. K	290.K	15	-	*

3.2 Output formats

3.2.1 Iso value contours

- Scale: 1 / 2 i.e. 20 cm represented by 10 cm
- Variables: Mach number, density: ρ/ρ_∞, C_p
- Increments:
  - $\Delta M = 0.25$ for problems 3.1, 3.2, 3.3.
  - $\Delta M = 0.5$ for problems 3.4, 3.5.
  - $\Delta M = 1.0$ for problem 3.6
  - $\Delta\rho/\rho_\infty = 0.1$
  - $\Delta C_p = 0.01$

3.2.2 Wall values

- Variables: C_p, C_f and St.
- Scales:
  - x axis: 1 / 2
  - y axis:

    $\Delta C_p = .25$ represented by 10 cm (2.5cm for test case 3.3)

    $\Delta C_f = 0.01$ represented by 10 cm (test cases 3.1 to 3.5)

$\Delta C_f = 0.005$ represented by 10 cm (test case 3.6)

St : use a logarithmic scale from $St = 10^{-5}$ to $St = 10^{-2}$ in which each decade is represented by 4cm.

3.2.3 Cross Flow Profiles

For the turbulent calculation (test case 3.3), profiles in the direction perpendicular to the wall (that is as functions of y) are required at x=0.15m and x=0.25m (hinge) from the leading edge. At x=0.15m, $\Delta y = 0.001m$ is represented by 2.cm and at $x = 0.25m$, $\Delta y = 0.01m$ is represented by 6.cm. Three profiles are required :

- velocity u/u_∞, with $\Delta(u/u_\infty) = 1$ represented by 10cm,
- temperature T/T_∞, with $\Delta(T/T_\infty) = 1$ represented by 2cm,
- shear stress $\tau_w/(0.5\rho_\infty u_\infty^2)$ with $\Delta 1000.\tau_w/(0.5\rho_\infty u_\infty^2) = 1$ represented by 5cm at x=0.15m and $\Delta 100.\tau_w/(0.5\rho_\infty u_\infty^2) = 1$ represented by 2.5cm at x=.25m.

4 Problem 4 : Flow over a 3-D Obstacle

4.1 Description of problem 4

Geometry:

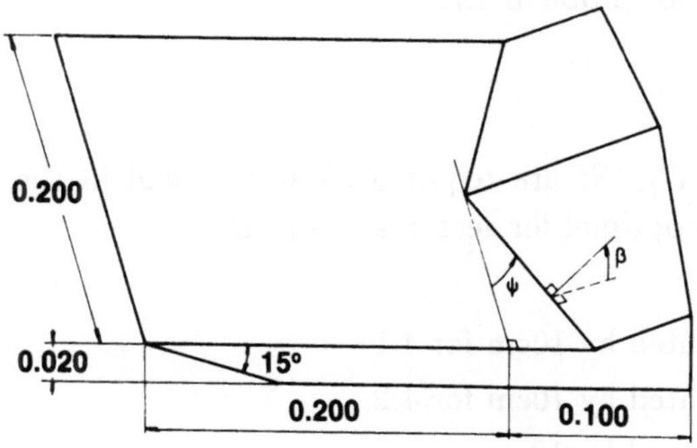

$\psi = 30^o$

Boundary conditions: Those of problem 1 plus appropriate lateral boundary conditions.

Test cases:

	M_∞	Re_∞/m	T_∞	T_{wall}	β	Modelization	Level
4.1	10.	$9.\ 10^6$	50. K	288.K	15	Lam. n. react.	* *
4.2	10.	$9.\ 10^6$	50. K	288.K	30	Lam. n. react.	* *

4.2 Output Formats

The obstacle is symmetric with respect to the Oxz plane (see the figure) so that the calculation can be restricted to the $y > 0$ part of the flow. The origin of the axes is the intersection point of the dihedron edge with the plate. All the dimensions are given in meters (notice the length of the flat plate before the ramp). Experimental measurements are made on the obstacle along lines located in 5 longitudinal planes :

- A : $y = 0$
- B : $y = 0.015$
- C : $y = 0.030$
- D : $y = 0.0475$
- E : $y = 0.065$

Heat fluxes are measured along A,C,D and static pressures along A,B,C,D,E.

4.2.1 Isovalue contours

Iso-Mach number and isobar contours are mandatory only in the symmetry plane (section A) and are optional in sections C and D.

- Scale : 1/2 i.e. 20 cm represented by 10cm.
- Variables : Mach number, pressure coefficient C_p. $\Delta M = .5$, $\Delta C_p = 0.01$ for 4.1, $\Delta C_p = 0.02$ for problem 4.2.

4.2.2 Wall values

The distributions of C_p, C_f, St are required along the wall in the symmetry plane (section A) and are optimal for sections C and D.

- Scale : Ox axis 1/2.
- $\Delta C_p = 0.25$ represented by 10cm for 4.1
- $\Delta C_p = 0.50$ represented by 10cm for 4.2
- $\Delta C_f = 0.01$ represented by 10cm
- St : use a logarightmic scale from $St = 10^{-5}$ to $St = 10^{-2}$ in which each decade is represented by 4cm.

4.2.3 Skin friction lines

In order to make at least a qualitative comparison with oil flow patterns, skin friction lines should be presented in projection on the $z = 0$ plane. Scale 1/2.

5 Problem 5: Corner Flow

5.1 Description of problem 5

Geometry:

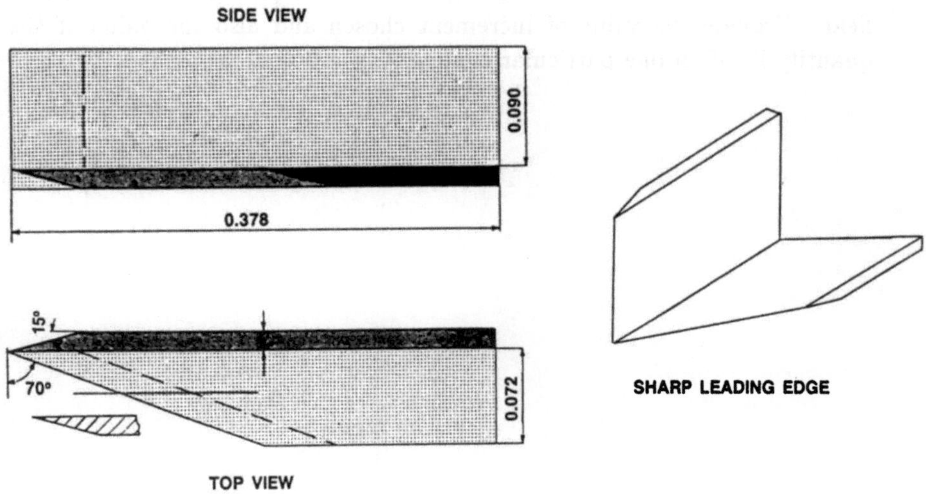

Boundary conditions: Those of problem 1 plus appropriate lateral boundary conditions.

Test case:
The angle of attack is 20. degree and the yaw angle is 0. degrees.

	M_∞	Re_∞/m	T_∞	T_{wall}	Modelization	Level
5.1	8.7	2.25 10^6	60. K	293.K	Lam. n. react.	****

Remark: it is possible to extend the geometry to apply numerical lateral boundary conditions.

5.2 Output formats

- **Locations and variables**
  - In the planes y=0 and z=0, iso-value contours of ρ/ρ_∞, C_p , C_f and St.
  - In the plane x = 0.2 m, and if possible, in the planes x = 0.1 m and x = 0.3 m :
    (a) iso-values contours of ρ/ρ_∞, C_p , T/T_∞ and Mach number.
    (b) field of cross-flow velocity vectors (y and z components).

- **Formats**
  - scale 48/100 for the planes $y = 0$ and $z = 0$, so that the iso-value contours of one quantity in these two planes can be grouped in one figure ;
  - scale 100/100 for the planes $x =$ Constant, so that two plots can be grouped in one figure.

Choose values of increments which give the best representation of the flow field. Provide the value of increment chosen and also the value of the quantity itself on one particular contour.

6 Problem 6: Double (Simple) Ellipsoid

6.1 Description of problem 6

Geometry:

The geometry is a double ellipsoid extended by a truncated cylinder. It is defined by:

$$x \leq 0. \qquad (\frac{x}{.06})^2 + (\frac{y}{.025})^2 + (\frac{z}{.015})^2 = 1$$

$$x \leq 0.\ z \geq 0. \qquad (\frac{x}{.035})^2 + (\frac{y}{.0175})^2 + (\frac{z}{.025})^2 = 1$$

$$0.0 \leq x \leq 0.016 \qquad (\frac{y}{.025})^2 + (\frac{z}{.015})^2 = 1$$

$$z \geq 0. \qquad (\frac{y}{.0175})^2 + (\frac{z}{.025})^2 = 1$$

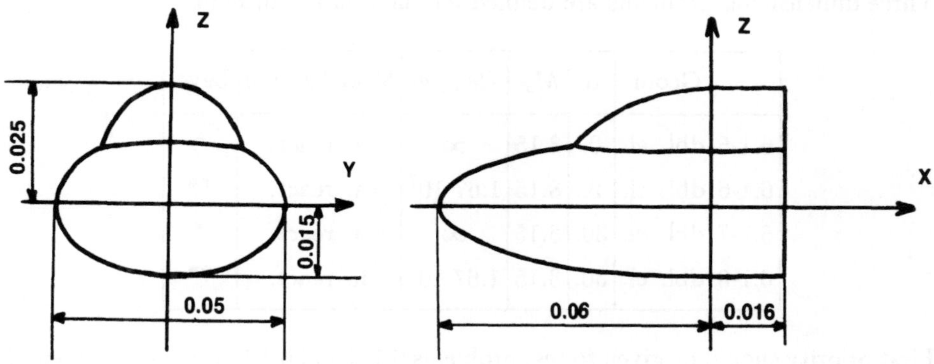

The geometry for the simple ellipse (ellipsoid) is provided by the previous equations. 2D calculations can be performed first on a simple or double ellipse obtained by taking $y = 0$.

6.2 Non reactive flows (problem 6.1) :

Boundary conditions:

- on the body:
  - $U_{wall} = 0$
  - $T_{wall} = 288K$ for cases 6.1-2 and 6.1-4.
- at infinity : Dirichlet boundary conditions are set according to the Mach number at infinity M_{∞}.

Remarks

- $T_{\infty} = 56\ K$.
- For Euler calculations, $Re = \infty$ and the only boundary condition on the wall is the slip boundary condition ($U_{wall} \cdot n = 0$)

- α is the angle of attack.
- For Navier Stokes calculations, the choice of the afterbody shape is left free.

Test cases (2D)

	Geom.	α	M_∞	Re_∞/m	Modelization	Level
6.1-1	dbl. el.	30.	8.15	∞	n. react.	*
6.1-2	dbl. el.	30.	8.15	1.67 10^7	n. react.	*
6.1-3	dbl. el.	30.	25.	∞	n. react.	*
6.1-4	dbl. el.	30.	25.	2.2 10^4	n. react.	*

Test cases (3D)
Three dimensional problems are defined for the double ellipsoid :

	Geom.	α	M_∞	Re_∞/m	Modelization	Level
6.1-5	dbl. el.	0.	8.15	∞	n. react.	*
6.1-6	dbl. el.	0.	8.15	1.67 10^7	n. react.	**
6.1-7	dbl. el.	30.	8.15	∞	n. react.	*
6.1-8	dbl. el.	30.	8.15	1.67 10^7	n. react.	**

First priority should be given to test problems 6.1-6 and 6.1-8 where experimental results are available.

6.3 Non-equilibrium reactive flows (problem 6.2) :

Model size: For test cases 6.2, the size of the body is scaled by a factor of 10. (semi-axis = .6 m).

Boundary conditions: Same as for cases 6.1 with the exception that $T_{wall} = 1500.K$.

Test cases (2D)

	Geom.	α	M_∞	Re_∞/m	Modelization	Level
6.2-1	simple el.	30.	25.	∞	react.	*
6.2-2	simple el.	30.	25.	2.2 10^4	react.	**
6.2-3	dbl. el.	30.	25.	∞	react.	**
6.2-4	dbl. el.	30.	25.	2.2 10^4	react.	**

(react. = reactive)

Test cases (3D)

Three dimensional problems are defined for the simple or double ellipsoid.

	Geom.	α	M_∞	Re_∞/m	Modelization	Level
6.2-5	simple el.	30.	25.	∞	react.	**
6.2-6	simple el.	30.	25.	$2.2\ 10^4$	react.	**
6.2-7	dbl. el.	30.	25.	∞	react.	***
6.2-8	dbl. el.	30.	25.	$2.2\ 10^4$	react.	***

Standard atmosphere is assumed (at 75km p=2.52 Pa, T=205.3K)

6.4 Equilibrium reactive flows (problem 6.3) :

Model size: For test cases 6.3, the size of the body is scaled by a factor of 10. (semi-axis = .6 m).

Boundary conditions: same as for cases 6.2.

Test cases (2D)

	Geom.	α	M_∞	Re_∞/m	Modelization	Level
6.3-1	simple el.	30.	25.	∞	$\gamma = 1.2$	*
6.3-2	simple el.	30.	25.	∞	react.	*
6.3-3	simple el.	30.	25.	$2.2\ 10^4$	react.	*
6.3-4	dbl. el.	30.	25.	∞	react.	*
6.3-5	dbl. el.	30.	25.	$2.2\ 10^4$	react.	*

Test cases (3D)

Three dimensional problems are defined for the simple or double ellipsoid.

	Geom.	α	M_∞	Re_∞/m	Modelization	Level
6.3-6	simple el.	30.	25.	∞	$\gamma = 1.2$	*
6.3-7	simple el.	30.	25.	∞	react.	*
6.3-8	simple el.	30.	25.	$2.2\ 10^4$	react.	**
6.3-9	dbl. el.	30.	25.	∞	react.	**
6.3-10	dbl. el.	30.	25.	$2.2\ 10^4$	react.	**

Standard atmosphere is assumed (at 75km p=2.52 Pa, T=205.3K)

6.5 Rarefied flows (problem 6.4) :

Model size: semi-axis = .06 m.

Boundary conditions: complete accomodation is assumed.

	Geom.	M_∞	Re_∞/m	λ_∞	T_{wall}	T_∞	Modelization
6.4-1	dbl. el.	25.	2.2 10^4	1.05 10^{-3}	620.	13.5	rarefied

The angle of attack is 30. degrees.
Level : ****

A three dimensional problem is defined for the double ellipsoid.

	Geom.	M_∞	Re_∞/m	λ_∞	T_{wall}	T_∞	Modelization
6.4-2	dbl. el.	25.	2.2 10^4	1.05 10^{-3}	620.	13.5	rarefied

The angle of attack is 30. degrees.
Level : ****

6.6 Output formats

6.6.1 Test cases 6.1 : Non-Reactive Flows

- **Iso-value contours**
  - Scale: 15 / 10 i.e. large semi-axis of 6 cm represented by 9 cm
  - Variables: Mach number, density: ρ/ρ_∞, C_p
  - Increments:
    - $\Delta M = 0.25$
    - $\Delta\rho/\rho_\infty = 0.25$
    - $\Delta C_p = 0.10$
  - For 3-D calculations, the field iso-value contours are requested in the planes x=0, y=0 and z=0, and the iso-value contours drawn on the body surface and projected onto these planes are also requested.

- **Wall values**
  - Variables: C_p, C_f, and St.
  - Scale:

- x axis: 15 / 10
- y axis:

  $\Delta C_p = 1.$ represented by 5 cm

  $\Delta C_f = 0.0150$ represented by 10 cm

  $\Delta St = 0.0075$ represented by 10 cm

- For 3-D calculations, the above outputs are requested in the plane $x = 0$, $y = 0$ and $z = 0$. To allow comparisons with oil-flow patterns, contributions should also include a side view of the wall streamlines or stress vectors.

6.6.2 Test cases 6.2 and 6.3: Reactive Flows

The x-axis and the z-axis are attached to the body and when they appear, they must be represented respectively horizontally and vertically on the plot.
Note : For Navier-Stokes calculations, the boundary condition $\frac{\partial Y}{\partial \eta} = 0$ can be applied over the fully non-catalytic wall.

- **(A) Iso-value contours**
  - Scale : 15/100 i.e. large semi-axis of 60 cm represented by 9 cm
  - Properties : $M = V/c$ (c : frozen sound speed), T and $C_p = \frac{p - p_\infty}{1/2 \rho_\infty V_\infty^2}$, mass fractions of products : Y_N, Y_O and Y_{NO} if visible.
  - Increments : $\Delta M = 0.25$, $\Delta T = 500$ K and $\Delta C_p = 0.10$, $\Delta Y = 0.01$.
- **(B) Wall values (lower and upper surfaces)**
  - Properties : M, T, C_p and the mass fractions Y_N, Y_O and Y_{NO} (if visible) as functions of x.
  - Scales :
    - x-axis : 15/100
    - M : $\Delta M = 1$ represented by 2.5 cm (for inviscid flows)
    - T : $\Delta T = 1000$ K represented by 1cm (for inviscid flows)
    - C_p : $\Delta C_p = 1$ represented by 5 cm ($C_p > 0$ represented downward)
    - Y : $\Delta Y = 0.10$ represented by 2.5 cm
    - C_f : $\Delta C_f = 0.2$ represented by 5 cm
    - St : $\Delta St = 0.5$ represented by 10 cm
- **(C) Properties along the stagnation streamline**
  - Properties : M, T and C_p as functions of the curvilinear distance from the stagnation point.
  - Scales :
    - vertically : identical to wall values, except for M : use $\Delta M = 2.5$
    - horizontally : free

- **(D) Three-dimensional calculations**

  - Same as the above, i.e. (A), (B) and (C), in the symmetry plane.
  - Iso-C_p contours in the plane $x = 0$: same specifications as in (A).
  - Diagram of the transversal Mach number

$$M_T = \frac{\sqrt{V_y^2 + V_z^2}}{c}$$

    (in which V_y and V_z are the y and z velocity components) evaluated on the body in the plane $x = 0$, as a function of y.
    Same scale as in (B).

6.6.3 Test case 6.4 : rarefied flows

- **General data**

  - Computer / memory size / maximal performance in MFLOP's.
  - Required CPU time / required memory.
  - CPU time to steady state / CPU time for averaging.
  - Gas collision model / gas surface interaction model.
  - Method of evaluation of the mean free path and/or Knudsen number. Explain how the modelled gas mean free path and its transport properties related to the real gas of the experiment.
  - Free stream number density.
  - Upstream boundary condition used (uniform or matched to wind tunnel flow ?)

- **Grid structure**

  - Details of number of cells and cell distribution
  - Provide a figure indicating the cell geometry

- **Flow-field results for plane of symmetry**
  Data is to be presented in the form of contour plots for the plane of symmetry.

- Physical scale × 1.5
- Iso-density, ρ/ρ_∞ of 0.05, 0.1, 0.25, 0.5, 0.75, 1.1, 1.5 and subsequently $\Delta\rho = 1.0$.
- Iso-Mach, $M = 0.5$ and subsequently $\Delta M = 1$.
- Iso-temperatures showing translational and internal temperatures separately. $\Delta T = 100$.

- **Flow-field results for a cross-flow plane** ; mandatory for x=0, others optional.
  Data is to be presented in the form of contour plots in format for the right hand half-plane only (Same format as in the symmetry plane).

- **Stagnation stream line results**
  - Scale : x axis, distance scaled × 6.
  - Density : $\rho/\rho_\infty = 12.0$ represented by 120 mm.
  - Translational temperature : $\Delta T = 2000$ K represented by 150 mm.
  - Internal mode temperature : $\Delta T = 2000$ K represented by 150 mm.

- **Surface fluxes**
  - iso-contours shown on the model surface

    i Physical scaled : × 1.5.
    ii. Stanton number : $\Delta St = 0.05$
    iii. Pressure coefficient : $\Delta C_p = 0.1$

  - Profiles along the plane of symmetry.

    i. Physical scale : × 1.5
    ii. Stanton number : 0 - 0.8 represented by 120 mm
    iii. Pressure coefficient : 0 to 2 represented by 100 mm
    iv. Skin friction coefficient : 0 to 2 represented by 100 mm

  - Skin friction lines on the surface.

- **Overall forces and heat transfer**

  Tabulated data for :

  - Lift coefficient.
  - Overall drag coefficient.

- Pressure drag coefficient.
- Shear force drag coefficient.
- Heat transfer coefficient.

- **Convergence history**

  Plot log RMS residual versus iterations or, for DSMC calculations, a representative function, e.g., overall drag, against simulated time and indicate averaging period for the presented mean values f parameters.

7 Problem 7: Flow over a Delta Wing

7.1 Description of problems 7

7.1.1 Laminar non reactive flows (problem 7.1):

Geometry:

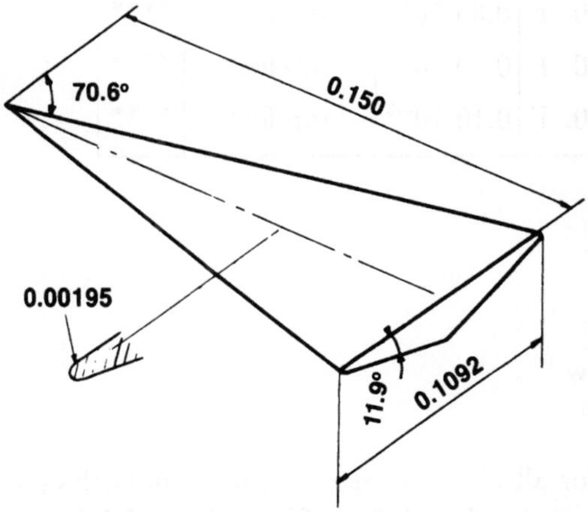

Boundary conditions: same as in Problem 1

Test Cases:

	M_∞	Re_∞/m	T_∞	T_{wall}	Modelization	Level
7.1-1	7.15	∞	74. K	-	n. react.	****
7.1-2	7.15	39. 10^6	74. K	288. K	lam. n. react.	****

The angle of attack is 30. degrees.

7.1.2 Rarefied gases (problem 7.2) :

Geometry:

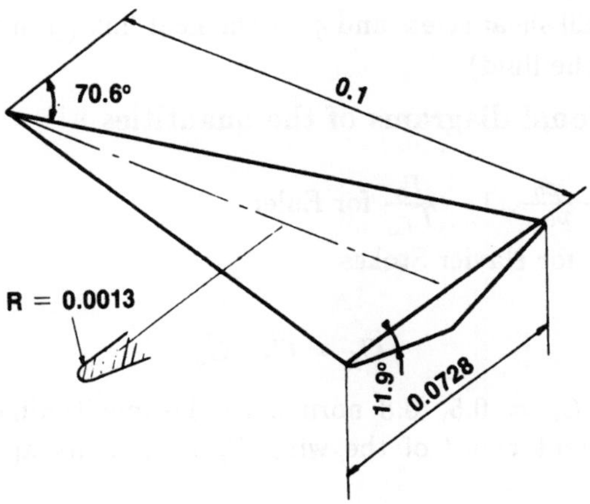

Boundary conditions: complete accomodation is assumed.

Test Cases:

	M_∞	Re_∞/m	T_∞	T_{wall}	λ_∞	Modelization	Level
7.2-1	20.2	2.8 10^4	13.5 K	620. K	1.05 10^{-3}	rarefied	****
7.2-2	20.	8.4 10^4	13.5 K	620. K	0.35 10^{-3}	rarefied	****
7.2-3	20.2	3.2 10^5	13.6 K	300. K	0.10 10^{-3}	rarefied	****
7.2-4	20.2	3.2 10^5	13.6 K	550. K	0.10 10^{-3}	rarefied	****

The angle of attack is 30. degrees.

7.2 Output formats

7.2.1 Problem 7.1 : Vortex flow

- **Global values**
  Give the values of C_D and C_L for all the test cases evaluated in each case by integration of the surface pressure. In addition for problem 7.1-2 give the values due to the integration of the skin friction.
- **Surface values**
  The surface coefficients for pressure, skin friction and Stanton number are defined respectively :

$$C_p = \frac{p - p_\infty}{\frac{1}{2}\rho_\infty {u_\infty}^2}$$

$$\vec{C}f = \frac{\vec{\tau}_w}{\frac{1}{2}\rho_\infty {u_\infty}^2}$$

$$St = \frac{\dot{q}_w}{\rho_\infty u_\infty (T_{0_\infty} - T_\omega)}$$

where $\vec{\tau}_\omega$ is the wall shear stress and $\dot{q}_\omega$ is the heat flux (positive if directed from the wall to the fluid).

  - **Cross-sectional diagrams of the quantities :**

C_p, M, $1 - \frac{p_0}{p_{0_\infty}}$, $1 - \frac{T_0}{T_{0_\infty}}$ for Euler
C_p, C_{f_y}, S_t for Navier Stokes
where

$$C_{f_y} = \vec{C}_f \, . \, \vec{C}_y$$

In planes $x/C_r = 0.5,\ 0.8$ normal to the longitudinal axis of the wing. The root chord of the wing C_r is from its apex. Plot the values vertically against (distance from the center-line)/(local gross semi-span) on a horizontal scale of 8 cm for one unit. The following vertical scales are likely to be suitable :

- C_p from $-.28.10^{-1}$ to $-.20.10^{-1}$; upper surface only, in 10 cm.
- M choose $\Delta M = 1$ for 1 cm ; in 10 cm ; Euler only
- $1 - \frac{p_0}{p_{0_\infty}}$ from 0 to 1 : in 10 cm ; Euler only
- $1 - \frac{T_0}{T_{0_\infty}}$ from -1.10^{-2} to $+2.10^{-2}$ in 10 cm ; Euler only
- C_{f_y} from -0.1 to 0.2 in 10 cm ; Navier-Stokes only
- St from -5.10^{-3} to 1.10^{-3} in 10 cm ; Navier-Stokes only.

– **Contour maps of the upper surface values of :**

- C_p, M, $1 - \frac{p_0}{p_{0_\infty}}$ for Euler
- C_p, S_t for Navier-Stokes

(as seen in projection on the horizontal plane $z = 0$ of the configuration) Use equal scales, in which the root chord of the wing is 12 cm. The following contour level intervals are likely to be convenient :

- $\Delta C_p = 5 \times 10^{-4}$
- $\Delta M = 0.2$
- $\Delta \left(1 - \frac{p_0}{p_{0_\infty}}\right) = 1 \times 10^{-3}$
- $\Delta S_t = 5 \times 10^{-5}$

– **Flow lines** Root chord of the wing is again 12 cm.
Surface streamlines for Problem 7.1-1.
This will show separation on upper surface. Give also the y-coordinate of separation at 50%.
Skin friction lines for problem 7.1-2.

• **Flowfield values**

– **Contour maps of :**

- C_p with $\Delta C_p = 5 \times 10^{-4}$ (upper part only i.e. $z \geq 0$)
- M $\Delta M = 0.2$
- $1 - \frac{p_0}{p_{0_\infty}}$ with $\Delta \left(1 - \frac{p_0}{p_{0_\infty}}\right) = 5 \times 10^{-3}$
- $1 - \frac{T_0}{T_{0_\infty}}$ with $\Delta \left(1 - \frac{T_0}{T_{0_\infty}}\right) = 5 \times 10^{-3}$
- $|\vec{\Omega}| = |\, \vec{curl}\, V\,|$, $\Delta\, |\vec{\Omega}|$ is not specified.
- $\frac{T}{T_\infty}$, $\Delta \left(\frac{T}{T_\infty}\right)$ not specified, Navier-Stokes only

Drawn in :

- **Symmetry plane** ($y = 0$) : C_p and M only. Root chord of the wing = 10 cm.

- **Cross sections** The above five quantities drawn in planes $x/c = .5$ and $x/c = .8$. Gross semi-span is 6 cm.

– **Velocity vectors :** Drawn in the cross-sectional planes $x/c = 0.5$ and 0.8. Use equal lateral and vertical scales in which the overall gross semi-span is 6 cm. See Fig.5.

– **Streamlines** Optional ; free formats.

- **Over the wing** To show the vortex.
- **On the upper surface** Surface streamlines for Euler, skin friction lines for Navier-Stokes.

– **Boundary layer profiles** Optional and free formats. Present some diagrams of the variation of density, velocity, temperature, and Mach number through the boundary layer on the upper surface in the planes $x/c = 0.5$ and 0.8.

- **Plot your mesh**
State your number of mesh points in each coordinate direction and indicate the size of the first layer of cells near body for Navier-Stokes. If mesh refinement have been done, summarize results.
- **Convergence history plot of the residual**
Use the most natural format. State your initial conditions.

7.2.2 Output formats for problem 7.2

- **General data**
  - Computer / memory size / maximal performance in MFLOP's.
  - Required CPU time / required memory.
  - CPU time to steady state / CPU time for averaging.
  - Gas collision model / gas surface interaction model.
  - Method of evaluation of the mean free path and/or Knudsen number. Explain how the modelled gas mean free path and its transport properties are related to the real gas of the experiment.
  - Free stream number density.
  - Upstream boundary condition used (uniform or matched to wind tunnel flow ?)
- **Grid structure**
  - Details of number of cells and cell distribution
  - Provide a figure indicating the cell geometry
- **Flow-field results for plane of symmetry**
Data is to be presented in the form of contour plots for the plane of symmetry and for cross flow planes (i.e. planes normal to the free stream direction). The latter is mandatory for x=80 mm for zero incidence and x

= 69.3 mm for 30° incidence (i.e. at 80% chord). For cross-flow plots show only right-hand half plane.

- Physical scale × 1.5
- Iso-density, ρ/ρ_∞ of 0.05, 0.1, 0.25, 0.5, 0.75, 1.1, 1.5 and subsequently $\Delta\rho/\rho_\infty = 1.0$.
- Iso-Mach, $M = 0.5$ and subsequently $\Delta M = 1$.
- Iso-temperatures showing translational and internal temperature separately. $\Delta T = 100$.

- **Stagnation stream line results**
  - Scale : x axis, distance scales × 10.
  - Density : $\rho/\rho_\infty = 3.5$ represented by 140 mm.
  - Translational temperature : $\Delta T = 12000K$ represented by 120 mm.
  - Internal mode temperature : $\Delta T = 12000K$ represented by 120 mm.
- **Surface fluxes**
  - Iso-contours shown on the model surface
    - **i.** Physical scale : × 1.5
    - **ii.** Stanton number $\Delta St = 0.05$
    - **iii.** Pressure coefficient : $\Delta C_p = 0.05$
  - Profiles along the symmetry plane and accross the wing at 80% chord. Other cross-flow planes can be presented but are not mandatory.
    - **i.** Physical scale : x axis 100 mm represented by 150 mm of plane of symmetry and 30 mm represented by 100 mm across wing. Show only the right-hand plane for the cross flow plot.
    - **ii.** Stanton number : Windward side 0.2 represented by 150 mm ; leeward side 0.01 represented by 150 mm.
    - **iii.** Pressure coefficient : -0.2 to 1 represented by 180 mm with an additional plot for the leeside with -0.3 to -0.1 represented by 100 mm.
  - Skin friction lines on the wing surface.
- **Overall forces**

  Plot results if data for different angles of attack are available, otherwise present tabulated results.
  - Incidence : x axis -5° to 45° represented by 120 mm.
  - Lift coefficient : -0.1 to 1 represented by 110 mm.
  - Overall drag coefficient : -0.1 to 1 represented by 110 mm.
  - Pressure drag coefficient : -0.1 to 1 represented by 110 mm.
  - Shear force drag coefficient : -0.1 to 1 represented by 110 mm.
  - Pitching moment coefficient : -0.05 to 0.2 represented by 100 mm.

- **Convergence history**
  Plot log RMS residual versus iterations or, for DSMC calculations, a representative function, e.g., overall drag, against simulated time and indicate averaging period for the presented mean values of parameters.

8 Problem 8: Non-equilibrium Flow in an Arc Jet or a Shock Tube

8.1 Description of problems 8

8.1.1 Non-equilibrium Flow in an Arc Jet (problem 8.1)

Geometry (axisymmetric):

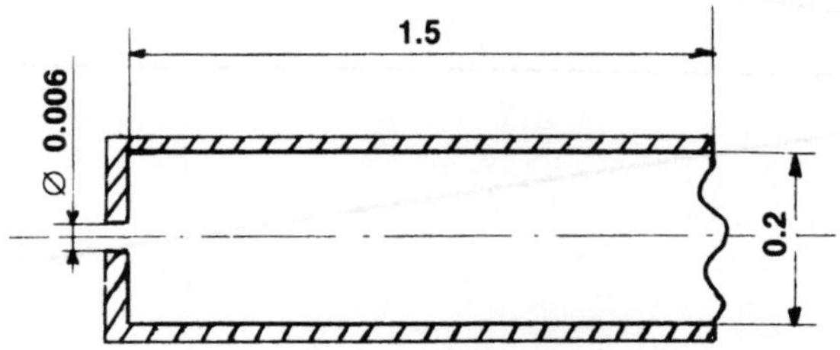

Physical conditions: For this test case, the flow is a free jet expanding into a plenum chamber. The jet is created by a torch whose diameter is equal to 0.006*m*. The starting condition will be given at the exit of the torch.

The gas considered is nitrogen. The species which are present are N_2^+, e, N, N^+, N_2. The pressure is equal to 250 tor into the plenum chamber. The total mass inflow is 50 Normal Litre/mn. The enthalpy is 13 MJoules/kg. The velocity profile is assumed to be parabolic.

The profiles of the rotational temperature (T_{rot}, taken equal to the translational temperature), the electronic temperature (T_e, assumed to be equal to the vibrational temperature) and concentrations of the different species are given from experimental results.

Level : ****

Experimental Profiles to be used as Starting Conditions

R (mm)	T_{el} (K)	T_{rot} (K)	E part/m^3	N part/m^3	N^+ part/m^3	N_2 part/m^3	N_2^+ part/m^3
0	11256	10264	$2.42\ 10^{22}$	$1.69\ 10^{23}$	$2.42\ 10^{22}$	$1.23\ 10^{22}$	$1.89\ 10^{19}$
.293	11200	9991	$2.34\ 10^{22}$	$1.71\ 10^{23}$	$2.34\ 10^{22}$	$1.74\ 10^{22}$	$1.89\ 10^{19}$
.568	11056	9445	$2.15\ 10^{22}$	$1.79\ 10^{23}$	$2.15\ 10^{22}$	$2.70\ 10^{22}$	$1.85\ 10^{19}$
.879	10781	8210	$1.80\ 10^{22}$	$1.90\ 10^{23}$	$1.80\ 10^{22}$	$5.90\ 10^{22}$	$1.83\ 10^{19}$
1.172	10407	6981	$1.39\ 10^{22}$	$2.06\ 10^{23}$	$1.38\ 10^{22}$	$1.01\ 10^{23}$	$1.81\ 10^{19}$
1.465	9970	6025	$9.84\ 10^{21}$	$2.22\ 10^{23}$	$9.82\ 10^{21}$	$1.47\ 10^{23}$	$1.76\ 10^{19}$
1.758	9429	5271	$6.11\ 10^{21}$	$2.42\ 10^{23}$	$6.09\ 10^{21}$	$1.93\ 10^{23}$	$1.73\ 10^{19}$
2.051	8829	4619	$3.31\ 10^{21}$	$2.60\ 10^{23}$	$3.29\ 10^{21}$	$2.47\ 10^{23}$	$1.66\ 10^{19}$
2.344 (wall)	8032	3957	$1.27\ 10^{21}$	$2.87\ 10^{23}$	$1.25\ 10^{21}$	$3.11\ 10^{23}$	$1.63\ 10^{19}$

(R : distance from axis)

8.2 Non-equilibrium Flow in a Shock Tube (problem 8.2)

Geometry (axisymmetric):

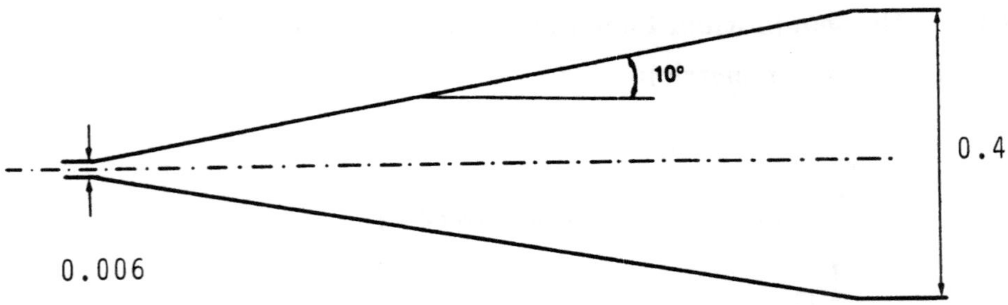

Divergent :

- half cone angle : $\Theta = 10^o$, length : $L = 1.13m$, outlet diameter : $D_{out} = .4m$, throat diameter : $D_{throat} = 6.10^{-3}m$.
- For the convenience of the computation a convergent part may be introduced upstream of the nozzle. For example, the divergent part may be completed upstream by a conical convergent :
- half cone angle : $\Theta = 45^o$, length : $l = 3.47\ 10^{-2}m$, inlet diameter : $D_{in} = 4.07\ 10^{-2}m$, throat diameter : $D_{throat} = 6.10^{-3}m$.

Physical conditions :

- The gas considered is air.
- reservoir conditions : nominal case : $P_0 = 1.53\ 10^8\ Pa$, $T_0 = 6500K$. These conditions correspond to a total enthalpy of approximately 10.55 MJ, and a mass flow rate of approximately 1.95 Kg/s.
- Participants are encouraged to present additional results for the more academic high temperature case : $P_0 = 10^8\ Pa$, $T_0 = 10000K$. They should use the scales defined hereafter except when these scales are inadequate.
- For participants starting their computation at the throat, Mach 1 and local equilibrium corresponding to reservoir conditions are assumed at the throat.
- wall conditions : $T_{wall} = 600K$, fully catalytic wall.
- Contributors should use Park chemistry model (Appendix 1) as basic model, and take into account vibrational relaxation of N_2 and O_2 (and possibly other diatomic species). Precaution should be taken when using Park's model for temperature below 700 K and equilibrium constants might need corrections. Coupling between vibration and dissociation may be considered. Several computational meshes and other chemical models may be tested, and for the high temperature case ionization may be included.
- Upstream conditions should be described.

- The participants are encouraged to clearly specify the physical models used and to describe the coupling between vibration and chemistry.

8.3 Output formats

8.3.1 Output formats for problem 8.1

- **Iso value contours in axial plane**
  - Axial Scale : 2 / 1 i.e. 10 cm represented by 20 cm
  - Radial Scale: 20 / 1 i.e. 0.3 cm represented by 6 cm
  - Variables: Rotational and Electronic temperatures: T_{rot},T_e
    Increments: ΔT = 1000. K from 3000. K upwards
- **Plots in radial planes**
  The plots are to be made in the radial planes defined by the following abscissa: X=0., 0.3, 0.8, 1.3, 1.8, 2.8 cm
  - The horizontal axis represents the radial axis of the flow.
  - Scale: 20 / 1 i.e. 0.3 cm represented by 6 cm
  - Variables to be plotted on the vertical axis : Rotational and Electronic temperatures, densities, velocity, pressure (T_{rot}, T_e, U, P, n_e, $n_{N_2^+}$, n_N, n_{N^+}, n_{N_2}).
  - Scales :
    - T : 2 cm for 1000. K from 3000. K upwards
    - U : 2 cm for 100. m/s
    - P : 1 cm for 25. Torrs from 250. Torrs upwards

  Note : Densities are in number of particles / cm^3.
    - A logarithmic scale will be used with 5cm for 1 decade.
    - n_e, n_{N^+} = ranging from 15 to 17 up to z = 0.8 cm, 14 to 16 for z larger than 0.8 cm.
    - $n_{N_2^+}$ = ranging from 12 to 14 for all
    - n_N, n_{N_2} = ranging from 16 to 18 for all

8.3.2 Output formats for problem 8.2

Introduction x axis is the symmetry axis, r axis is the radial axis of the nozzle. The origin of x axis is the nozzle throat, the origin of r axis is the symmetry axis. Since the nozzle is axisymmetric, values will be reported only on the upper half part of the nozzle.

In the following, M refers to frozen Mach number, P to static pressure, and T to static (translational) temperature. When not specified otherwise logarithm means logarithm of base 10. Frozen Mach number is defined by : $M = \frac{u}{c_f}$ where u is the fluid velocity and c_f the frozen speed of sound. T_ν is the vibrational temperature, s the specific entropy. Mass fraction of species A is represented by Y_A, concentration by [A]. Wall heat flux noted q is obtained by : $q = -\chi \cdot \nabla T \cdot \vec{n}$, $\vec{n}$ being the normal to the surface (wall), ∇T the gradient of temperature and χ the heat conductivity of the mixture. This heat conductivity may be obtained taking Prandtl number of the mixture equal to one.

Boundary layer displacement thikness δ_1 depends on an external velocity u_e and external density ρ_e. These external values of properties will be taken equal to the value of the same property on the axis for the same value of x. This displacement thickness will be computed by : $\delta_1 = \int_0^\delta (1 - \frac{\rho u}{\rho_e u_e})dr$, δ being the boundary layer thickness obtained using the same convention for external velocity.

In order to distinguish between equilibrium non-equilibrium and frozen regimes, characteristic times for vibration τ_ν, dissociation τ_d and recombination τ_r, and for convection τ_c are introduced. They are defined as follows (ignoring diffusion) :

Let us consider vibrational energy equation :

$$\frac{dE^i_\nu}{dt} = S^i$$

where E^i_ν is the vibrational energy of species i and S^i a source term for vibrational energy that may include coupling between vibration and dissociation and Landau Teller term. A characteristic time for vibration is :

$$\tau^i_\nu = \frac{E^i_v}{S^i}$$

Similarly, when writing species continuity equation (without diffusion, for sake of clarity) :

$$\frac{d[A^i]}{dt} = R^i - D^i$$

where $[A^i]$ is the concentration of the diatomic species A^i, R^i the source term corresponding to recombinations and D^i the source term corresponding to dissociations. The following time scales can be introduced : $\tau^i_d = \frac{[A^i]}{D^i}$ and $\tau^i_r = \frac{[A^i]}{R^i}$ and finally, convective time scale τ_c : $\tau_c = L/\nu$. where L is the nozzle length and ν the velocity of the fluid.

- **Iso-values contours :**
  - scale : 1/15 for the nozzle, 1/2 for the enlargement of the throat region $0 < x < 0.3m$ when such an enlargement is useful.
  - properties :
    - Frozen Mach number M
    - $T, T_\nu(O_2), T_\nu(N_2)$
    - Y_{O_2}, Y_{N_2}, Y_{NO}

- specific entropy s
- increments : $\Delta M = .25, \Delta T = \Delta T_\nu = 200.K, \Delta Y_{O_2} = \Delta Y_{NO} = 5.10^{-3} \Delta Y_{N_2} = 1.10^{-3}$
- entropy : free. Participants should specify the scale they use.

- **Properties along symmetry axis**
  - frozen Mach number M
  - $T, T_\nu(0_2), T_\nu(N_2)$ on the same plot
  - Y_{O_2}, Y_N, Y_{NO} (three plots)
  - specific entropy s
  - τ_c, τ_ν, τ_d and τ_r on two different plots : one for N_2 and one for O_2 .
  - pressure P
  - **scales :** Contributors should use logarithmic scales for Y_N, P and characteristic time scales τ
    - scale of horizontal axis : 1/15 i.e. 1 cm represents 15 cm
    - vertical axis :
      - $\Delta M = 1$ corresponds to 1cm
      - $\Delta T = 500K$ corresponds to 1 cm .
      - $\Delta Y_{O_2} = 10^{-2}$ and $\Delta Y_{NO} = 10^{-2}$ correspond to 1 cm. $\Delta(log Y_N) = 2$ corresponds to 1 cm.
      - log scales : 1 cm per unit increment of log except for $log Y_N$.

- **Properties along the wall**
  - heat flux q
  - boundary layer displacement thickness δ_1
  - scale of horizontal axis : 1/15 i.e. 1 cm represents 15 cm
  - vertical axes : scales free.

- **Profiles**

  These profiles will be plotted for two stations : $x = 0$ (throat) and $x = 1.13m$ (exit). The properties will be plotted with respect to the nondimensional parameter $\eta = \frac{(r_{wall} - r)}{r_{wall}}$.
  The value of this parameter is 0 at the wall and 1 at the axis of symmetry. η will vary along the vertical axis of the plot, the physical properties along the horizontal axis

  plots :

  - velocity v
  - temperatures $T, T_\nu(O_2), T_\nu(N_2)$ on the same plot
  - mass fractions Y_{O_2}, Y_N, Y_{NO}

- scales : vertical axis : 10 cm for $\Delta\nu = 1$; vertical axis : 2 cm for $\Delta T = 500K$ $\Delta y_{O_2} = 10^{-2}$ and $\Delta Y_{NO} = 10^{-2}$ correspond to 1 cm. $\Delta(log Y_N) = 2$ corresponds to 1 cm.
- velocity scale : at the nozzle throat, $\Delta\nu = 250m/s$ for 1 cm. at the nozzle exit : $\Delta\nu = 500m/s$ for 1 cm .

- **Global properties**
  - logarithm of residual (L2 norm) versus number of iterations
  - mass flow rate Q through each grid section as a function of x.
  - scale : horizontal : x, 1 cm for 15 cm length (1/15 scale). number of iterations : free
  - vertical : 1 cm for a variation of the logarithm of residual equal to 1. $\Delta Q = 10^{-2}$ Kg/s corresponds to 1 cm.

Appendix: Non-equilibrium model for Problem 6

A- Reactional scheme

The reactional scheme is the 5-species, 17-reaction model by PARK [1]

1. **Indices** :
   $i = 1, \cdots, 5$ index of species M_i ($M_i = O,\ N,\ NO,\ O_2$ or N_2)
   $r = 1, \cdots, 17$ index of chemical reaction R_r
2. **Reaction** R_r :

$$\sum_{i=i_l} \sigma^l_{i,r}\, M_i \rightleftharpoons \sum_{i=i_r} \sigma^r_{i,r}\, M_i$$

   $\sigma^l_{i,r}$, $\sigma^r_{i,r}$ stoichiometric coefficients
   i_l, i_r values assumed by i respectively on the left or right side of reaction R_r
3. **Constants** :

$$\begin{cases} K^{equ}_r(T) = \exp(A_1 + A_2 Z + A_3 Z^2 + A_4 Z^3 + A_5 Z^4) \\ K^{frd}_r(T) = C\,T^s \exp(-D/kT) \\ K^{rev}_r(T) = K^{frd}_r / K^{equ}_r \end{cases}$$

   in which K^{equ}_r (resp. K^{frd}_r and K^{rev}_r) are the equilibrium constant (resp. forward reaction constant and reverse reaction constant) of reaction r, $Z = 10\ 000/T$ (T in Kelvin), and the constants A_1, A_2, A_3, A_4, A_5, C, s, and D/k are given in the table (p. 248).
4. **Production Rates** :

$$\Omega_i = m_i \sum_{r=1}^{17} \left(\sigma^r_{i,r} - \sigma^l_{i,r}\right) J_r$$

   in which m_i is the molar mass of species M_i, and

$$J_r = K^{frd}_r(T) \prod_{i=i_l} \left(\frac{\rho_i}{m_i}\right)^{\sigma^l_{i,r}} - K^{rev}_r(T) \prod_{i=i_r} \left(\frac{\rho_i}{m_i}\right)^{\sigma^r_{i,r}}$$

B- Governing equations for inviscid reactive flow

1. **Species conservation** : (non-equilibrium flow)

$$(\rho_i)_t + \text{div}\,(\rho_i \mathbf{V}) = \Omega_i \qquad (i = 1, \cdots, 5)$$

[1] C. PARK, "On Convergence of Chemically Reacting Flows", AIAA Paper 85-0247, AIAA 23rd Aerospace Sciences Meeting, Jan. 14-17, 1985/Reno, Nevada.

($\mathbf{V}$: velocity vector; $\rho_i = \rho Y_i$: density of species M_i; Y_i: mass fraction of species M_i).

2. **Momentum** : Usual form.
3. **Energy Equation** :

$$E_t + \text{div}\,(\mathbf{V}\,(E+p)) = 0$$

where $E = \rho(h + \frac{V^2}{2}) - p$ (total energy per unit volume), and h (specific enthalpy) is taken to be:

$$h = \sum_{i=1}^{5} Y_i C_{p_i} T + \sum_{i=1}^{3} Y_i h_i^0 + \sum_{i=3}^{5} Y_i e_i^{vib}$$

in which the indices from 1 to 5 refer to O, N, NO, O_2 and N_2 respectively,

$$C_{p_i} = \begin{cases} (5/2)\,R_i & \text{for i=1 and 2} \\ (7/2)\,R_i & \text{for i=3, 4 and 5} \end{cases}$$

($R_i = \mathcal{R}\,/\,m_i$: gas constant of species M_i; $\mathcal{R}$ = 8.3144 J.K^{-1}.mole^{-1}), the formation heats of N and O can be related to the dissociation temperatures of O_2 and N_2,

$$\begin{cases} h_1^0 = h_O^0 = R_1\,\Theta_{O_2}^d/2 \\ h_2^0 = h_N^0 = R_2\,\Theta_{N_2}^d/2 \end{cases}$$

and [2]

$$\begin{cases} \Theta_{O_2}^d = 59\ 500\ \ \text{K} \\ \Theta_{N_2}^d = 113\ 000\ \ \text{K} \end{cases}$$

the formation heat of NO is

$$h_3^0 = h_{NO}^0 = 2.99\ 10^6\ \ \text{J.kg}^{-1}$$

and e_i^{vib} designates the vibrational energy of the diatomic species M_i.

4. **Vibrational Energy** : For equilibrium-flow computations, use the following expression

$$e_i^{vib} = \frac{R_i\,\Theta^i{}_v}{\exp\left(\Theta^i{}_v/T\right) - 1}$$

where Θ_i^v is the characteristic vibrational temperature of the molecular species M_i ** :

$$\begin{cases} \Theta_3^v = \Theta_{NO}^v = 2\ 740\ \ \text{K} \\ \Theta_4^v = \Theta_{O_2}^v = 2\ 270\ \ \text{K} \\ \Theta_5^v = \Theta_{N_2}^v = 3\ 390\ \ \text{K} \end{cases}$$

[2] W. G. VINCENTI and C. H. KRUGER, Jr., Introduction to Physical Gas Dynamics, R. E. Krieger Publishing Co., Malabar/Florida, Inc., 1982, p. 524.

For non-equilibrium flow computations, to facilitate the comparison of the performance of the numerical methods when applied to a non-equilibrium model that contains the equilibrium model as a limiting case, contributors are requested to also employ the above equilibrium vibrational energy although not physically correct. However, contributors were strongly encouraged to also perform computations using more sophisticated non-equilibrium models.

5. **State Equation** : Assume a mixture of perfect gases :

$$p = \sum_{i=1}^{5} \rho_i R_i T = \rho \mathcal{R} T \sum_{i=1}^{5} \frac{Y_i}{m_i}$$

C- Governing equations for viscous reactive flows :

For viscous non-equilibrium flow computations, entire freedom was left to contributors to enhance the inviscid model (B) by physically relevant extensions; also, if results for more complete reactional schemes are presented, computations with the reactional scheme described in A are nevertheless required to permit the comparison of numerical procedures.

APPENDIX I

Initial species	Final species	Reaction energy kcal/mole	Equilibrium constant coefficients A1	A2	A3	A4	A5	Rate constant cm3/mole C	Temperature exponent S	Activation temperature D/K
					Dissociation					
1. 02 + M =	0 + 0 + M	-117,98	1.335	-4.127	-0.616	0.093	-0.005			
							M = N	8.250E + 19	-1.000	59500.
							M = 0	8.250E + 19	-1.000	59500.
							M = N2	2.750E + 19	-1.000	59500.
							M = 02	2.750E + 19	-1.000	59500.
							M = NO	2.750E + 19	-1.000	59500.
2. N2 + M =	N + N + M	-225.00	3.898	-12.611	0.683	-0.118	0.006			
							M = N	1.110E + 22	-1.600	113200.
							M = 0	1.110E + 22	-1.600	113200.
							M = N2	3.700E + 21	-1.600	113200.
							M = 02	3.700E + 21	-1.600	113200.
							M = NO	3.700E + 21	-1.600	113200.
3. NO + M =	N + 0 + M	-150.03	1.549	-7.784	0.228	-0.043	0.002			
							M = N	4.600E + 17	-0.500	75500.
							M = 0	4.600E + 17	-0.500	75500.
							M = N2	2.300E + 17	-0.500	75500.
							M = 02	2.300E + 17	-0.500	75500.
							M = NO	2.300E + 17	-0.500	75500.
					Exchange reactions					
4. NO + 0 =	N + 02	-32.05	0.215	-3.6576	0.843	-0.136	0.007	2.160E + 08	1.290	19220.
5. 0 + N2 =	N + NO	-74.97	2.349	-4.828	0.455	-0.075	0.004	3.180E + 13	0.100	37700.

A. V. Johansson, P. H. Alfredsson (Eds.)

Advances in Turbulence 3

Proceedings of the Third European Turbulence Conference, Stockholm, July 3–6, 1990

1991. XI, 540 pp. 366 figs. Hardcover ISBN 3-540-53595-0

The book covers the following main topics: turbulence structure, transition, dynamical systems in relation to transition, turbulent combustion and mixing, turbulence affected by body forces, turbulence modeling, drag reduction, and novel experimental techniques. In addition to 6 invited survey lectures the reader will find well over 50 contributions by specialists in the field.
The book addresses researchers in physics, engineering, and numerical simulation.

H.-H. Fernholz, H. E. Fiedler (Eds.)

Advances in Turbulence 2

Proceedings of the Second European Turbulence Conference, Berlin, August 30–September 2, 1988

1989. XII, 522 pp. 392 figs. Hardcover ISBN 3-540-51274-8

This volume contains six invited lectures and many research papers presented by scientists and engineers from different disciplines actively engaged in turbulence research, both experimental and theoretical. The scope of the meeting encompassed free and wall-bounded turbulent shear flows, computational methods and novel measuring techniques.
Special emphasis was laid on fundamental aspects of the following topics: stability and generation of turbulence, effects of rotation, stratification and buoyancy forces, novel instrumentation, manipulation and control, boundary layers with separation and reattachment, computer simulation, turbulent diffusion, image analysis and flow visualization, vorticity dynamics and turbulence, and large-scale structures.
The book will be of particular interest to scientists working in fluid mechanics, meteorology and applied mathematics.

G. Comte-Bellot, J. Mathieu (Eds.)

Advances in Turbulence 1

Proceedings of the First European Turbulence Conference, Lyon, France, July 1–4, 1986

1987. XVI, 586 pp. 437 figs. Hardcover
ISBN 3-540-17586-5

MIX
Papier aus verantwortungsvollen Quellen
Paper from responsible sources
FSC® C105338

If you have any concerns about our products,
you can contact us on
ProductSafety@springernature.com

In case Publisher is established outside the EU,
the EU authorized representative is:
Springer Nature Customer Service Center GmbH
Europaplatz 3, 69115 Heidelberg, Germany

Printed by Libri Plureos GmbH
in Hamburg, Germany